Praise for

Dan Chiras

When you need practical advice from a warm,
smart and informed human being, Dan Chiras is
the one to turn to.

— BRUCE KING, PE Director,
Ecological Building Network; and author,
Buildings of Earth and Straw and *Making Better Concrete*

Dan Chiras has done as much as anyone in America
to promote and popularize the use of renewable energy.

— STEPHEN MORRIS, publisher and editor,
Green Living: A Practical Journal for the Environment;
and editor, *The New Village Green: Living Light,
Living Local, Living Large*

Dan Chiras is one of the most authoritative writers
in the field of renewable energy. His multiple books create
a comprehensive library for homeowners looking to live
a lifestyle in harmony with their values.

— DAVID JOHNSTON, *What's Working:
Visionary Solutions for Green Building*; and author,
Green Remodeling: Changing the World One Room at a Time

power
from the
sun
achieving energy
independence

power
from the
sun achieving energy
independence

DAN CHIRAS

with Robert Aram
and Kurt Nelson

NEW SOCIETY PUBLISHERS

CATALOGING IN PUBLICATION DATA:
A catalog record for this publication is available from the National Library of Canada.

Cover design by Diane McIntosh.
Cover images: House: © iStock/Andreas Weber; Wildflowers — © iStock/Chen Ping-Hung

Printed in Canada by Friesens.
First printing August 2009.

Paperback ISBN: 978-0-86571-621-6

Inquiries regarding requests to reprint all or part of *Power from the Sun* should be addressed to New Society Publishers at the address below.

To order directly from the publishers, please call toll-free (North America) 1-800-567-6772, or order online at www.newsociety.com

Any other inquiries can be directed by mail to:

New Society Publishers
P.O. Box 189, Gabriola Island, BC V0R 1X0, Canada
(250) 247-9737

New Society Publishers' mission is to publish books that contribute in fundamental ways to building an ecologically sustainable and just society, and to do so with the least possible impact on the environment, in a manner that models this vision. We are committed to doing this not just through education, but through action. This book is one step toward ending global deforestation and climate change. It is printed on Forest Stewardship Council-certified acid-free paper that is **100% post-consumer recycled** (100% old growth forest-free), processed chlorine free, and printed with vegetable-based, low-VOC inks, with covers produced using FSC-certified stock. Additionally, New Society purchases carbon offsets based on an annual audit, operating with a carbon-neutral footprint. For further information, or to browse our full list of books and purchase securely, visit our website at: **www.newsociety.com**

NEW SOCIETY PUBLISHERS

Mixed Sources
Cert no. SW-COC-001271
© 1996 FSC

FSC

Books for Wiser Living
recommended by *Mother Earth News*

Today, more than ever before, our society is seeking ways to live more conscientiously. To help bring you the very best inspiration and information about greener, more sustainable lifestyles, *Mother Earth News* is recommending select New Society Publishers' books to its readers. For more than 30 years, *Mother Earth* has been North America's "Original Guide to Living Wisely," creating books and magazines for people with a passion for self-reliance and a desire to live in harmony with nature. Across the countryside and in our cities, New Society Publishers and *Mother Earth* are leading the way to a wiser, more sustainable world.

Contents

CHAPTER 1

AN INTRODUCTION TO
SOLAR ELECTRICITY

Many people dream of a world powered by solar energy. They envision solar-powered cars, solar-powered homes, and even solar factories. It may come as a surprise to many readers to learn that the Sun is already the main source of energy in our world; it powers nearly all forms of life — from the simplest single-celled organisms to the most complex plants and animals, even humans.

Solar energy makes its way to us via photosynthetic algae and plants. These organisms form the base of both aquatic and terrestrial food chains. They use sunlight energy to produce food molecules via photosynthesis. During this process, solar energy is trapped in the chemical bonds that link the atoms in these molecules. Food molecules, in turn, are consumed directly, as we humans dine on vegetables, or are passed up the food chain from one organism to another, making its way to other animals that we consume as

well. The energy in the food molecules in the plants and animals we consume is released in an intricately choreographed series of chemical reactions in the cells of our bodies. It is then used to power cells, including muscle and nerve cells. Solar energy powers every move we make and every breath we take. As you turn the pages of this book, you're using solar energy captured by plants.

Solar energy's contribution to global energy supplies doesn't end there, however. All fossil fuels contain massive amounts of solar energy. How did solar energy end up in coal and oil and fuels like gasoline derived from oil? Consider coal first.

Coal is ancient plant matter that once grew in and along the banks of lakes, streams and coastal swamps many millions of years ago. Just as modern plants species do, these ancient species captured the Sun's energy and used it to create all the organic molecules they required to grow and reproduce. As plants died or shed leaves, massive amounts of solar-energy-rich organic matter was deposited on the bottoms of these aquatic ecosystems. There, they were buried by sediments eroded from soil in local watersheds. Over time, pressure from the thick layers of sediment and heat from the Earth's interior converted the organic waste into peat. Continued heat and pressure transformed the peat into coal.

Today, we extract this coal and burn it in power plants to generate electricity. During combustion, the energy-rich chemical bonds of the ancient plant matter in coal are broken down. When they break, the chemical bonds release ancient solar energy stored for millions of years. This energy powers our computers and lights up our lives.

The energy we unleash from oil is also ancient solar energy. Oil deposits, however, were formed not from plants or dinosaurs, as is commonly thought, but from much tinier members of the living world, single-celled marine algae. Like plants, these algae captured sunlight energy and stored it in the chemical bonds of the molecules they produced via photosynthesis. In ancient times, trillions upon trillions of these microscopic organisms sank to the bottom of the ocean, accumulating in thick deposits on the sea floor. They, too, were covered by sediment. Over time, pressure from the overlying sediment and heat from the Earth's interior slowly converted this organic material into the thick gooey material we call crude oil. Oil, in turn, is refined to produce gasoline, jet fuel, and diesel, which are burned in vehicles, releasing ancient solar energy. It powers our trucks, cars, buses, jets, and airplanes.

The energy released by the combustion of natural gas is also derived from ancient solar

Where Does Your Electricity Come From?

Did you ever wonder how your local utility generates the electricity you use to power your home or business and what the source of that energy is? According to the US Energy Information Administration, nationwide electricity sold by utilities comes from the sources shown in the accompanying table.

As in many cases, national averages can be deceiving. Your utility may rely almost entirely on coal, nuclear, or hydropower. You can find out by calling your utility and talking with their public information specialists.

Coal	52%
Nuclear	21%
Natural Gas	16%
Hydro	8%
Other Renewables	2%
Petroleum	1%

Source: Energy Information Administration, "Net Generation by Energy Source, 2006 Data for Electric Utilities and Independent Power Producers," www.eia.doe.gov.

energy. Natural gas is a by-product of the formation of oil and coal. That is to say, it is one of many chemicals formed when organic matter in ancient plants and algae was transformed into coal and oil millions of years ago. This gaseous by-product is often extracted from coal and oil deposits. So, just like coal and oil, the energy liberated when natural gas burns comes from sunlight that fell on the Earth many millions of years ago.

Even hydropower, the energy captured by flowing water, is a form of solar energy. Sunlight striking the Earth's surface causes evaporation. Water that evaporates from the sea, land, and surface waters accumulates in the atmosphere and eventually falls to Earth as rain and snow that feed streams and rivers. Water flowing downstream by gravity in rivers and streams, in turn, powers turbines that generate electricity.

Solar energy is also responsible for generating winds. This secondhand solar energy can be tapped to pump water and make electricity.

Although we live in a solar-powered world, the ancient solar energy that powers our homes, businesses, and vehicles is not what many of us have in mind when we envision a renewable energy future. Ancient solar resources such as oil, natural gas, and coal contain hydro-carbons and other chemicals. When these fuels are burned, they release millions of tons of harmful pollutants. Some pollutants are altering our climate; others are poisoning our cities and towns, and still others are raining down on us in the form of harmful acids that cause untold damage to buildings, crops, and ecosystems.

Solar energy advocates dream of clean, reliable, and affordable solar energy that will serve all of our energy needs. We envision solar power that heats our homes, the water in which we bathe, and even the food we eat. Solar energy could provide electricity to power lights, electronic devices, and appliances — even our cars and buses. In the process, it could sideline or minimize the solar-charged, though polluting, fossil fuel resources whose use is wreaking havoc on people, the economy, and the environment.

This book focuses on an important portion of the solar dream: the production of electricity by solar electric systems, also known as *photovoltaic systems* or *PV systems* for short. In this book, we'll focus primarily on smaller PV systems for homes and businesses — systems whose rated power typically ranges from 1,000 to 10,000 watts.

An Overview of Solar Systems

As their name implies, solar electric systems convert sunlight energy into electricity. This conversion takes place in the solar modules, also commonly referred to as solar panels. A solar module consists of numerous solar cells (Figure 1-1). Solar cells are made from one of the most abundant chemical substances on Earth, silicon — a component of sand and quartz that makes up much of Earth's crust.

Solar cells are wired together and encased in plastic and glass to create solar modules. The plastic and glass layers protect the solar cells from the elements, especially moisture. Two or more modules are typically mounted on a rack and wired together. Together, the

Rated Power and Capacity

Solar homeowners and solar installers typically refer to their solar electric systems in terms of rated power, also known as rated capacity. A homeowner, for instance, might tell you that she has a PV system rated at 3,000 watts. *Rated power* is the instantaneous output of a solar module measured in watts under standard test conditions (STC). Watts is a measure of power. It is the rate of flow of energy and is used to rate technologies that produce electricity, such as solar cells, and consumers of electricity, such as light bulbs. We'll describe watts in more detail shortly, but remember that it is an instantaneous measure of the output of an electric-generating device or consumption of a device that consumes electricity.

The rated output of a PV module is determined by flashing light at 1,000 watts per square meter, which is equivalent to full sun, on a module that is maintained at 77°F (25°C) — only slightly warm. In this procedure, the Sun's rays arrive perpendicular to the solar module. (Light rays striking the module perpendicular to the surface result in the greatest absorption of sunlight.) Rating modules under standard test conditions provides a rating that buyers can refer to when comparing one module to another from the same and different manufacturers. It is also used when sizing a solar system.

Most residential solar electric systems fall within the range of 1,000 to 6,000 watts or 1 to 6 kilowatts (kW) (1,000 watts is equal to 1 kilowatt). It is important to note, however, that a 3 kW PV system won't produce 3,000 watts of electricity all the time the Sun's shining on it. It only produces this amount under standard test conditions. So, this rating system leaves a bit to be desired.

When mounted outdoors, PV modules typically operate at higher temperatures than those that occur under standard test conditions. That's because infrared radiation in sunlight striking PV modules causes them to heat up. To achieve a cell temperature of 77°F (25°C) under full sun, like that measured in the laboratory, the air temperature must be quite low — about 23° to 32°F (0° to -5°C), not a typical temperature for PV modules in most locations most of the year. This is significant because higher cell temperatures decrease the efficiency of PV cells.

To better simulate real-world conditions, the solar industry has developed an alternative rating system. They call it *PTC*, which stands for *PV-USA Test Conditions* (which stands for *PV for Utility Scale Applications Test Conditions*).

PTC were developed at the PV-USA test site in Davis, California and more closely approximate real-life operating conditions. In this rating system, the ambient temperature is held at 68°F (20°C). The modules face the Sun and output is measured when the Sun's irradiance reaches 1,000 watts per square meter. The conditions also take into account the cooling effect of wind. Wind speed in the test is 2.24 miles per hour (or 1 meter per second) 10 meters above the surface of the ground. (That is, wind speed is monitored at 10 meters, or 33 feet.) Although the cell temperature varies with different modules, it is typically about 113°F (45°C) under PTC. ☞

In general, PTC ratings are roughly 10% lower than STC ratings, because power output decreases with rising temperature. It decreases about 0.5% for every 1°C increase in temperature. Because there are differences in PV modules, however, the ratio of PTC to STC can vary dramatically.

When shopping for PV modules, we suggest that you ignore the nameplate rating the manufacturers provide and look up the PTC ratings. Some manufacturers publish this data on their spec sheets; others don't. You can look up PTC ratings on the California Energy Commission's website: www.consumerenergycenter.org. ■

rack and solar modules are referred to as an *array*.

Electricity generated by a PV array flows via wires to yet another component of the system, the *inverter*. It converts the direct current electricity produced by solar cells into the alternating current electricity commonly used in our homes and businesses.

Applications

Once a curiosity, solar electric systems are becoming commonplace, even in less-than-sunny areas such as Germany and Japan. (Germany and Japan are the largest markets for PV modules in the world today.) While many solar electric systems are being installed around the world to provide electricity to homes, they are also being installed on schools, small businesses, and office buildings, even skyscrapers like 4 Times Square in New York City, home of NASDAQ. Many large corporations such as Microsoft, Toyota, and Google have also installed large solar electric systems. Some electric utilities are also installing

Fig. 1-1: *Solar-electric systems generally consist of two or more modules wired together to form an array. Each module consists of multiple solar cells wired in series to increase the voltage.*

large PV arrays to supplement conventional sources.

Even colleges are getting in on the act. Colorado College, where Dan teaches courses on renewable energy and global warming, has installed a solar system to offset its electrical demand. The Madison Area Technical College in Wisconsin has also installed a solar electric system. Some airports are installing PV systems to meet their needs. Sacramento's airport, for instance, installed PVs over parking structures to shade vehicles and generate electricity; Denver International Airport installed a 2,000 kilowatt solar electric system in 2008. On a smaller scale, ranchers often install solar electric systems to power electric fences to contain their livestock or to pump water into remote stock-watering tanks.

Solar electricity is proving to be extremely useful on boats. Sailboats, for example, are often equipped with solar electric systems — usually only a few hundred watts — to power lights, fans, radio communications, GPS systems, and refrigerators. Many recreational vehicles (RVs) are equipped with small PV systems that are used to power microwaves, TVs, and satellite receivers. The electricity these systems produce reduces the run time for gas-powered generators that can disturb fellow "campers."

Many bus stops and parking lots are illuminated by solar electricity, as are portable information signs used at construction sites. Numerous police departments now haul solar-powered radar units to neighborhoods to discourage speeding. These units display a car's speed and warn drivers if they're exceeding the speed limit.

Backpackers and river runners can even take roll-up solar modules with them on their ventures into the wild to power electronics. Dan's son carries a portable solar electric charger for his iPod, which, much to his father's chagrin, he listens to as the two of them backpack through the Rockies (Figure 1-2). Dan uses the same charger throughout the rest of the year to recharge his cell phone.

Fig. 1-2: This small solar device is designed to charge portable electronic equipment, including cell phones, PDAs, and other small devices such as IPods.

DAN CHIRAS

Solar electric systems are well suited for remote applications where it is too costly to run power lines. Dan designed a solar system for a client to power a small organic farm in rural Canada to supply a water pump, refrigerator, and remote security camera. PV systems supply electricity to many remote homes, cottages, and cabins. In France, the government paid to install solar electric systems and wind turbines on farms at the base of the Pyrenees Mountains, rather than running electric lines to these distant operations.

Emergency call boxes found in many remote stretches of highway in North America are powered by solar electricity, as are many highway warning lights (Figure 1-3). Solar electricity is also used to power remote monitoring stations that collect data on rainfall, temperature, and seismic activity. Stream flow monitors on many US rivers and streams rely on solar-powered transmitters to beam data to solar-powered satellites. The data is then

Fig. 1-3: *Solar modules are the product of choice in remote locations. Solar-powered emergency call boxes like the one shown here on a remote highway in northern California are now found along rural highways in many states.*

Power and Energy: What's the Difference?

The terms "power" and "energy" are frequently used, but often misunderstood. In the electrical world, power is measured in watts or kilowatts. Power is an instantaneous measure, like the speed of a car. In the electrical world, power (watts) is a measure of the rate of flow of energy. Engineers and scientists rate electrical loads, devices that consume electricity such as light bulbs and electric motors, in watts or kilowatts. For example, a light bulb might be rated at 100 watts. An electric motor might be rated at 1,000 watts. Scientists and engineers also use watts, as noted in a previous sidebar, to rate electric-generating technologies such as PV systems. A solar system, for instance, might be rated at 3,000 watts.

Energy, in contrast, is power consumption or production over time. A light bulb that consumes 100 watts for one hour, consumes 100 watt-hours of electricity. If it operates for 10 hours, it uses 1,000 watt-hours of energy or 1 kilowatt-hour of energy. Both watt-hours and kilowatt-hours are measures of energy use. A solar electric system producing electricity *at a rate of* 1,000 watts for one hour produces 1,000 watt-hours or 1 kilowatt-hour of energy.

beamed back to Earth to the US Geological Survey, where it is processed and disseminated. PVs also allow scientists to gather and transmit data back to their labs from remote sites, like the tropical rain forests of Central and South America.

Solar electric modules often power lights on buoys, which are vital for nighttime navigation on large rivers like the Saint Lawrence Seaway. Railroad signals and aircraft warning beacons are also often solar powered.

PV modules are used to boost radio, television, and telephone signals. Signals from these sources are often transmitted over long distances. For successful transmission, however, they must be periodically amplified at relay towers. The towers are often situated in inaccessible locations, far from power lines. Because they are reliable and require little, if any, maintenance, PV systems are ideal for such applications. They make it possible for us to communicate across long distances. Next time you make a long-distance telephone call from a phone on a landline, rest assured solar energy is making it possible.

While PV systems are becoming very popular in more developed countries, they're also widely used in less developed nations. They are, for instance, being installed in remote villages in less developed countries to power lights and televisions, and refrigerators and freezers used to store vaccines and other medicines. They're also used to power pumps to provide water for villages.

The ultimate in remote and mobile applications, however, has to be the satellite. Virtually all military and telecommunications satellites are powered by solar electricity, as is the International Space Station.

World Solar Energy Resources

Although solar electricity is growing in popularity, it provides a tiny fraction of today's electrical demand. However, as global supplies of fossil fuel resources decline and as concern over global climate change increases, solar electric systems could become a major source of electricity, along with wind systems and other renewable energy technologies. But is there enough solar energy to meet our needs?

Although solar energy is not evenly distributed over the Earth's surface, significant resources are found on every continent. "Solar energy's potential is off the chart," write energy experts Ken Zweibel, James Mason, and Vasilis Fthenakis in a December 2007 article, "A Solar Grand Plan" published in *Scientific American* magazine. Less than one billionth of the Sun's energy strikes the Earth, but, as they point out, the solar energy striking the Earth in a 40-minute period is equal to all the energy human society consumes in a year. Solar electric systems mounted on our homes and businesses or giant commercial solar systems could tap into the Sun's generous supply of energy, providing us with an abundance of electricity.

Could solar electric provide 100% of the United States' or the world's electrical energy needs?

Yes, it could.

Will it?

Probably not.

In fact, no one is planning on a 100% solar future. Rather, most renewable energy experts envision a system that consists of a mix of renewable energy technologies such as solar hot water, solar electricity, passive solar, wind, hydro, geothermal, and biomass combined with radical improvements in energy efficiency. In areas where wind is abundant, for example, it could become a major source of electricity.

In a sustainable energy system, PV systems could produce enormous amounts of electricity for homes, businesses, farms, ranches, schools, and factories. Another solar technology, large-scale solar thermal electric systems, could supplement PV systems and play a huge role in meeting our needs. Solar thermal electric systems concentrate sunlight energy to generate heat that's used to boil water. Steam generated from this process is used to spin a turbine connected to a generator that makes electricity (Figure 1-4). Some of the newest solar thermal electric systems even store hot water so electricity can be generated on cloudy days or during the evening.

Wind turbines could also provide a significant amount of electricity to power our future, perhaps even more than PVs. Geothermal and biomass resources could contribute as well. *Biomass* is plant matter that can either be burned directly to produce heat to generate steam that's used to make electricity or converted to liquid or gaseous fuels that can be burned to produce electricity or heat. Hydropower will continue to contribute to the energy mix in a renewable energy future. What will happen to conventional fuels such as oil, natural gas, coal (burned as cleanly as

a

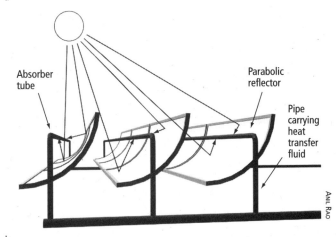

b

Fig. 1-4: *(a) Solar thermal electric systems like the one shown here produce electricity at cost-competitive rates. (b) Sunlight is focused on a pipe that carries a heat transfer fluid. The heat is transferred to water at a central point, causing it to boil. Steam produced by the boiling water drives a turbine. The spinning turbine is attached to a generator that produces electricity.*

possible), and nuclear energy? Although their role could diminish over time, these fuels will be part of the mix for many years to come. In the future, they could be used to back up renewable energy technologies that generate electricity.

Despite what critics say, solar energy is an abundant resource and many forms of solar energy capture technologies are affordable right now. Solar energy will very likely play a major role in our energy future. It has to. Fossil fuels like oil are finite. In fact, conventional oil (crude oil, not shale oil) could be economically depleted within 30 to 50 years. Many energy experts believe that crude oil supplies are already on the decline. (This is partly responsible for the record high gasoline prices in 2007 and much of 2008.) Natural gas supplies are on the decline, too. In fact, US natural gas production peaked in 1973. Some energy experts believe that global natural gas production could peak between 2015 and 2025. Coal, which is abundant, is still a finite resource and, lest we forget, a major source of the greenhouse gas carbon dioxide, the main cause of global warming.

The Sun, on the other hand, will continue to shine for at least 5 billion more years. With 30 to 50 years of oil left and 5 billion years of sunlight, what's the future going to look like?

What the Critics Say

Solar energy is a seemingly ideal fuel source. It's clean. It's free. It's abundant. Its use could ease many of the world's most pressing problems such as global climate change. Like all fuels, solar energy is not perfect. Critics point out that, unlike conventional resources such as coal, the Sun is not available 24 hours a day. Some people don't like the looks of solar electric systems. And solar electric systems are pretty pricey, too. Let's take a look at these arguments and respond to the criticisms.

Availability and Variability

Although the Sun shines 24 hours a day and beams down on the Earth at all times, half the planet is always in darkness. This poses a problem, as modern societies consume electricity 24 hours a day, 365 days a year.

Another problem is the variability of solar energy. That is, even during daylight hours clouds can block the sun, sometimes for days on end. If PV systems are unable to generate electricity 24 hours a day like coal-fired and nuclear power plants, how can we use them to power our 24-hour-per-day demand for electricity?

Homeowners like Dan who live off-grid (not connected to the electrical grid) solve the problem by installing batteries that store electricity to meet their nighttime demand and to supply electricity for use on cloudy days. As a result, Dan and folks like him are supplied with electricity 24 hours a day, 365 days a year by PV systems.

The sun's variable nature can also be offset by coupling solar electric systems with other renewable energy sources, for example, wind-electric systems, or micro hydro systems. Wind systems, for instance, generate electricity day and night — so long as the winds blow. Micro hydro systems tap the energy of flowing water in streams or rivers. Either one can be used to generate electricity to supplement a PV system, compensating for the Sun's "shortcomings." Solar and wind are especially good partners. Figure 1-5 shows data on the solar and wind resources at Dan's renewable energy education center in eastern Missouri. As illustrated, the Sun shines a lot in the spring, summer, and early fall but less so during the winter. During winter, however, the winds blow more often and blow much more forcefully. A wind turbine could easily make up for the reduced output of a PV system, ensuring a reliable, year-round supply of electricity at this and other similar sites.

On an individual level, then, the Sun's shortcomings are easily overcome. But can society find a way to meet its 24-hour-per-day needs for electrical energy from the Sun?

Possibly.

Scientists and engineers are currently developing numerous ingenious technologies to store solar electricity. Batteries are not high on the list, however. Why?

To store massive amounts of electricity to power factories, stores, and homes, we'd need massive battery banks. Because they would be costly and would require huge investments, scientists are seeking a variety of other, potentially more cost-effective options. One option is the solar thermal electric system mentioned earlier, which stores surplus hot water to run generators at night or during cloudy periods.

Another option is the use of solar electricity to power air compressors. They'd produce compressed air that could be stored in abandoned underground mines. When electricity is needed, the compressed air would

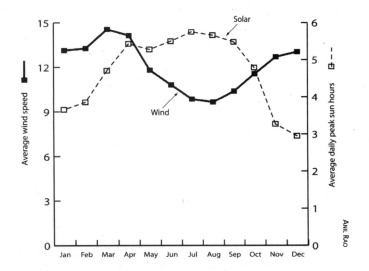

Fig. 1-5: *This graph of solar and wind energy resources at The Evergreen Institute's Center for Renewable Energy and Green Building in east central Missouri (operated by Dan Chiras) illustrates how complementary wind and sun are. As illustrated, the sunlight is fairly plentiful in the spring, summer, and fall, but not the winter. Wind picks up in the fall, winter, and early spring. A hybrid solar-electric/wind system can tap into these resources, providing an abundance of electricity year round.*

be released through a turbine, not unlike those found in conventional power plants. The blades of the turbine would be attached to a shaft that is attached to a generator that produces electricity.

Surplus solar electricity could also be used to generate hydrogen gas from water. Hydrogen gas is created when electricity is run through water. This process, known as electrolysis, splits the water into its components, hydrogen and oxygen, both gases. Hydrogen can be stored in tanks and later burned to produce hot air or to heat water to produce steam. Hot air and steam can be used to spin a turbine attached to a generator. Hydrogen could also be fed into a fuel cell, which produces electricity.

As in residential systems, electricity for mass consumption can also be supplied on cloudy days or at night by another renewable energy resource, the wind. That is, commercial wind farms could provide power to supplement solar systems because the winds often blow when the Sun is behind clouds

Fig. 1-6: *The author's boys, Skyler and Forrest, check out a large wind turbine at a wind farm in Canestota, New York. Wind farms like this one are popping up across the nation, indeed across the world, producing clean, renewable electricity to power our future.*

(during storms, for instance). Other renewable energy technologies can also be used to complement solar energy on a commercial scale. Hydroelectric plants and biomass facilities, for instance, could be used to ensure a continuous supply of renewable energy in a system that's finely tuned to switch from one energy source to another. In Canada, hydroelectric facilities are treated as a *dispatchable energy resource* much like natural gas is today. That is, they can be turned on and off as needed to meet demand. Such systems could be used to supplement electrical demand when it exceeds the capacity of commercial solar systems or at night.

Shortfalls could also be offset on a local or regional level by transferring electricity from areas of surplus solar and/or wind production to areas of insufficient electrical production. Surplus solar-generated electricity from Colorado, for example, could be shipped via the electrical grid to neighboring Wyoming, New Mexico, and Nebraska when needed. Surplus electricity generated on wind farms in Kansas could be shipped to neighboring Colorado, Nebraska, and Oklahoma (Figure 1-6).

Solar's variable nature can also be offset by natural gas-fired power plants and newer coal-fired plants that burn pulverized coal. Both can be started or stopped, or throttled up or down, to provide additional electricity. These facilities could serve as an excellent backup source as we transition to a renewable energy future.

There's no debate: solar energy is a variable resource. It's only available during the

Integrating Renewables into America's Electrical Grid

Although European nations such as Denmark, Germany, and Spain are successfully integrating renewable energy into their electrical grids, the electrical grid in the United States is not finely tuned enough to switch from one energy source to another. At times, the grid can hardly manage existing conventional sources. The size of the United States and the fact that the existing grid is overtaxed make integrating renewable energy more difficult. In Europe, shorter distances and a more robust electrical grid make it easier. Also, there is strong political and popular support for renewable energy in Europe.

daylight hours and, even then, can be reduced by cloud cover. Even though solar energy is not available 24 hours a day, there are ways to overcome this shortcoming. That said, we should point out that even though solar energy is variable, it is not unreliable. Just like wind energy, you can count on a certain amount of solar energy each year. With smart planning and careful design, we can meet a good portion of our electrical needs from this seemingly capricious resource.

Aesthetics

While many of us view a solar electric array as a thing of great value, even beauty, some don't. Your neighbors, for instance, might think that a solar electric system detracts from the beauty of the neighborhood. Because not everyone views a PV system the same way, some neighborhood associations have banned PV systems.

Ironically, those who object to solar electric systems rarely complain about the visual blight in our environs, among them cell phone towers, water towers, electric transmission

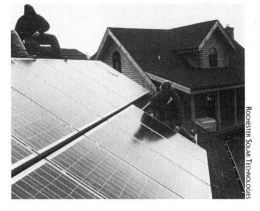

Fig. 1-7: *Solar arrays can be flush mounted, that is, mounted close to the roof to reduce their visibility. The arrays are typically mounted on racks three to six inches from the surface of the roof which helps cool the modules. While aesthetically appealing, flush-mounted arrays often produce less electricity than a rack-mounted array because the output of PV modules typically decreases at higher temperatures.*

ROCHESTER SOLAR TECHNOLOGIES

lines, and billboards. One reason that these common eyesores draw little attention is that they have been present in our communities for decades. We've grown used to ubiquitous electric lines and radio towers.

Fortunately, there are ways to mount a solar array so that it blends seamlessly with the roof. As you'll learn in Chapter 9, solar modules can be flush mounted on roofs (Figure 1-7).[1] There's also a solar product that can be applied directly to certain types of metal roofs, creating an even lower-profile array (Figure 1-8). Solar arrays can also be

Fig. 1-8: *Energy Conversion Devices/Uni-Solar produces a solar laminate, a thin-film solar product, that can be applied directly to standing seam metal roofs on homes, but more commonly on businesses.*

mounted on a pole or rack anchored to the ground that can be placed in sunny backyards out of a neighbor's line of sight.

Cost

Perhaps the biggest disadvantage of solar electric systems is that they're costly — very costly. Although the cost of solar cells has fallen precipitously, from around $50 per watt in the mid-1970s to $5 per watt in the early 2000s, solar electric systems are one of the most expensive means of generating electricity on the planet — but only if you ignore the environmental costs of conventional power and the generous subsidies these technologies receive from taxpayers.

Although solar electric systems are expensive, there are ways to lower the cost — often substantially. And there are factors that make a system competitive with conventional electricity. If, for instance, you live in an area with lots of sunshine and high electrical rates, such as southern California or Hawaii, a PV system competes very well with conventional electricity. Even in areas with low sunshine but high electrical rates, such as Germany, PV is economically competitive. Financial incentives for PV systems from local utilities or the state and federal government can drive costs down, often making solar electricity cost competitive with conventional sources.

If you are building a home more than a few tenths of a mile from a power line, solar electricity can also compete with utility power. That's because utility companies often charge customers a hefty fee to connect to the utility grid. You could, for instance, pay $20,000 to connect to the electric line, even if you're only a few tenths of a mile from a power line. Line extension fees don't pay for a single kilowatt-hour of electricity; they only cover the cost of the transformer, poles, wires, electrical meter, and installation. You'll pay the cost of the connection either up front or pro-rated for many years to come.

The Advantages of Solar Electric Systems

Although solar electricity, like any fuel, has some downsides, they're clearly not insurmountable and are outweighed by their advantages. One of the most important advantages is that solar energy is an abundant and renewable resource. While natural gas, oil, coal and nuclear fuels are limited and on the decline, solar energy will be available to us as long as the Sun continues to shine — for at least 5 billion years.

Solar energy is a clean resource, too. By reducing our reliance on coal-fired power plants, solar electricity could help homeowners and businesses reduce their contribution to a host of environmental problems, among them acid rain, global climate change, habitat destruction, and species extinction. Solar electricity could even replace costly, environmentally risky nuclear power plants. Although solar electricity does have its impacts, it is a relatively benign technology compared to fossil fuel and nuclear power plants.

Solar energy could help us decrease our reliance on declining and costly supplies of fossil fuels like natural gas. Solar could also help us decrease our reliance on oil. (Although very little electricity in the United States comes from oil, electricity generated by solar electric systems could someday be used to power electric or plug-in hybrid cars and trucks [Figure 1-9]). And, although the production of solar electric systems does have its impacts, all in all it is a relatively benign technology compared to fossil fuel and nuclear power plants.

Another benefit of solar electricity is that, unlike oil, coal, and nuclear energy, the fuel is free. Moreover, solar energy is not owned or controlled by hostile foreign states or one of

Fig. 1-9: *Plug-in hybrids like the one shown here and electric cars (not shown) will very likely play a huge role in personal transportation in the future. Electric cars with longer-range battery banks could be used for commuting and for short trips, under 200 miles, while plug-in hybrids could be used for commuting and long-distance trips.*

CAL CARS

the dozen or so major energy companies that dictate energy policy, especially in the United States. Because the fuel is free and will remain free, solar energy can provide a hedge against inflation, caused in part by ever-increasing fuel costs.

An increasing reliance on solar and wind energy could also ease political tensions worldwide. Solar and other renewable energy resources could alleviate the perceived need for costly military operations aimed at stabilizing the Mideast, a region where the largest oil reserves reside. Because the Sun is not owned or controlled by the Middle East, we'll never fight a war over solar or other renewable energy resources. Not a drop of human blood will be shed to ensure the steady supply of solar energy to fuel our economy.

Yet another advantage of solar-generated electricity is that it uses existing infrastructure — the electrical grid, and technologies in use today such as electric toasters, microwaves, and the like. A transition to solar electricity could occur fairly seamlessly.

Solar electricity is also modular. You can add on to a system over time. If you can only afford a small system, you can start small and expand your system as money becomes available.

Solar electricity could provide substantial economic benefits for local, state, and regional economies. And solar electricity does not require extensive use of water, an increasing problem for coal, nuclear, and gas-fired power plants, particularly in the western US and in arid regions.

Purpose of this Book

This book focuses primarily on small solar electric systems suitable for homes and small businesses. This book is written for individuals who aren't particularly well versed in electricity and electronics. That is, you won't need a degree in electrical engineering or physics to understand this material.

In this book, we have strived to explain facts and concepts clearly and accurately, introducing key terms and concepts as needed, and repeating them as often as prudent to make our points clearly and accurately. Our overarching goal was to create a user-friendly book.

When you finish reading and studying the material in this book, you'll know an amazing

Solar Energy and the Price of Conventional Energy

Relying on solar and other renewable energy resources could protect us from price gouging by foreign cartels and multinational corporations. It could also protect us from rising prices caused by declining supplies, increasing demand from rapidly industrializing countries such as China and India, and a decline in the value of the US dollar. Bear in mind, however, that even though solar energy is a free resource, the technologies that allow us to capture it are not. These systems require energy to manufacture, so the cost of solar electric systems are subject to rising prices of fossil fuels.

amount about solar energy and solar electric energy systems. You will have the knowledge required to assess your electrical consumption and the solar resource at your site. You will also be able to determine if a solar electric system will meet your needs and if it makes sense for you. You will have a good working knowledge of the key components of solar electric systems, especially PV modules, racks, controllers, batteries, and inverters. In keeping with our long-standing goal of creating savvy and knowledgeable buyers, this book will help you know what to look for when shopping for a PV system. You'll also know how PV systems are installed and what their maintenance requirements are.

We should point out, however, that this book is not an installation manual. When you're done reading, you won't be qualified to install a solar electric system — that usually requires a few hands-on workshops and competent advice and assistance. Even so, this book is a good start. You'll understand much of what's needed to install a system or even launch a new career in PVs. If you choose to hire a solar energy professional to install a system (a route we highly recommend) you'll be thankful you've read and studied the material in this book. The more you know, the more informed input you will have into your system design, components, siting, and installation — and the more likely you'll be happy with your purchase.

This book should also help you develop realistic expectations. We believe that those interested in installing renewable energy systems need to proceed with eyes wide open.

Knowing the shortcomings and pitfalls of solar electric systems (or any renewable energy technology, for that matter) helps us avoid mistakes and prevents disappointments that are often fueled by unrealistic expectations.

Organization of this Book

Now that you know a little bit about the ways solar electricity is being used and the pros and cons of this clean, renewable energy source, and the purpose of this book, let's begin our exploration of solar electricity. Here's how we'll proceed: We'll begin in the next chapter by studying the Sun and the energy it produces. We will discuss important terms and concepts such as peak sun hours. You'll learn why days (daylight hours) are longer in the summer and shorter in the winter and why the Sun's angle changes — and what implications the changing angles have on solar electric systems. You'll learn about *altitude angle*, *azimuth angle*, and the all-important *tilt angle*.

In Chapter 3, we'll explore solar electricity — the history of PVs, the types of solar cells on the market today, how solar cells generate electricity, and what new solar electric technologies are being developed.

In Chapter 4, we will explore the feasibility of tapping into solar energy to produce electricity at your site. We'll teach you how to determine your electrical energy needs and how to determine if your site has enough solar electricity potential to meet them. You'll also learn why energy-efficiency measures are so important and how they can reduce your initial system cost.

In Chapter 5, we'll examine three types of residential solar electric systems: (1) off-grid, (2) batteryless grid-tie, and (3) grid-connected with battery backup. You'll learn the basic components of each system. We'll also examine hybrid systems — for example, wind-PV systems — and why they can provide a reliable, year-round supply of electricity. You'll learn about net metering and ways to make solar electricity affordable. We'll explore how easy it is to expand a solar electric system and examine basic system design.

Chapter 6 introduces you to inverters, vital components of solar electric systems. You'll learn what they do and how they operate. We'll also provide some shopping tips — ideas on what you should look for when buying an inverter.

In Chapter 7, we'll tackle storage batteries and controllers, key components of off-grid solar systems or grid-connected systems with battery backup. You will learn about the types of batteries you can install, how to care for batteries, and ways to reduce battery maintenance. We will point out common mistakes people make with their batteries and give you ways to avoid making those same mistakes. You will also learn about battery safety and how to size a battery bank for a solar electric system. We'll finish with a discussion of charge controllers, looking at the way they work and why they are important in battery-based systems. We'll also explore backup generators, providing information on what your options are and what to look for when buying a generator.

In Chapter 8, we'll provide an overview of solar electric system installation and maintenance. You'll learn how to size a PV system and various mounting options — for example, pole-mounts and roof racks. We'll explore pole-mounted racks that enable PV arrays to track the Sun from sunrise to sunset and discuss the economics of this more efficient option.

In Chapter 9, we'll explore a range of issues such as permits, covenants, and utility interconnection. We'll discuss whether you should install a system yourself or hire a professional and, if you choose the latter, how to locate a competent installer.

Finally, as in all of Dan's books, this book includes a resource guide. It contains a list of books, articles, videos, associations, organizations, workshops, and websites of manufacturers of the components of PV systems.

CHAPTER 2

UNDERSTANDING THE SUN AND SOLAR ENERGY

The Sun lies in the center of our solar system, approximately 93 million miles from Earth. Composed primarily of hydrogen and small amounts of helium, the Sun is a massive nuclear reactor. But it's not the same type of reactor found in the nuclear power plants that many utility companies use to generate electricity. Those are *fission* reactors designed to split atoms of uranium-235 in a controlled fashion. The heat generated in this process boils water in a reactor to generate steam, which spins a turbine that drives a generator to make electricity.

The Sun is a giant *fusion* reactor. The fusion occurs in the Sun's core, where intense pressure and heat force hydrogen atoms to fuse, creating slightly larger helium atoms. In this process, immense amounts of energy are released; this energy migrates to the surface of the Sun, and then radiates out into space, primarily as light and heat.

Solar radiation streaming into space strikes the Earth, warming and lighting our planet, and fueling aquatic and terrestrial ecosystems. What's remarkable, though, is that the Earth receives only a tiny fraction of the Sun's output — about one half of one billionth of the energy that radiates from its surface. Although our allotment is small, that tiny fraction is equal to 170 million gigawatts of power. According to French energy expert, Jean-Marc Jancovici, "the solar energy received each year by the Earth is roughly ... 10,000 times the total energy consumed by humanity." To replace *all* the oil, coal, gas, and uranium currently used to power human society with solar energy, we'd need to capture a mere 0.01% of the energy of the sunlight striking the Earth each day. As mentioned in Chapter 1, 40 minutes worth of sunlight is equivalent to all the energy consumed by human society in a year!

19

According to the US Department of Energy's National Renewable Energy Laboratory (NREL), to generate the electricity the United States consumes, which represents 40% our nation's total energy consumption, we'd only need to install PVs on 7% of the total land surface area currently occupied by cities and homes. We could achieve this by installing PVs on rooftops, over parking lots, and on the sides of buildings.

Solar Variation

The sun's intensity varies by region. In the United States, for instance, the US Southwest is blessed with sunlight. On an annual basis, however, Kansas City receives only about 25% less sunlight than Phoenix. Buffalo, in one of the cloudiest regions in the United States, receives about 50% less sunlight than Kansas City. Solar electricity works in all regions, although the size of systems must be increased in cloudier regions to make up for reduced sunshine.

PVs and Solar Absorption

The vast majority of solar cells in use today are crystal silicon based — that is, they are made from silicon in crystal form. As noted in the text, these cells produce electricity from energy in the visible and lower end of the infrared portion of the electromagnetic spectrum — radiation in the 300 to 1,100 nanometer range. (A nanometer [nm] is one billionth of a meter.) Thin-film solar modules generally respond to a much narrower range — from 300 to slightly under 600 nm — although newer designs can increase the range to over 1,100 nm.

In other words, we wouldn't have to appropriate a single acre of new land to make PV our primary source of electricity.

In this chapter, we'll explore the Sun and solar radiation. We'll examine key concepts vital to solar electric systems: irradiance, irradiation, insolation, and peak sun hours. We'll examine the difference between direct and diffuse radiation and explore key concepts such as azimuth angle, altitude angle, and tilt angle. An understanding of these concepts is vital for installers. The last three are especially important for mounting PV arrays for optimum annual output.

Understanding Solar Radiation

The Sun releases immense amounts of energy from its surface every day. This output, known as *solar radiation*, consists primarily of electromagnetic radiation.[1] As shown in Figure 2-1, the Sun's electromagnetic radiation ranges from high-energy, short-wavelength gamma rays to low-energy, long-wave radiation known as radio waves. In between these extremes, starting from the short-wave end of the spectrum are x-rays, ultraviolet radiation, visible light, and heat (infrared radiation).

While the Sun releases numerous forms of energy, most of it (about 40%) is infrared radiation (heat) and visible light (about 55%). Traveling at a speed of 186,000 miles per second, solar energy takes a little over 8 minutes — 8.3 minutes, to be more precise — to make its 93 million mile journey from the Sun to the Earth. PV modules capture the energy contained in the visible and lower end of the infrared portions of the spectrum.

Electromagnetic Spectrum

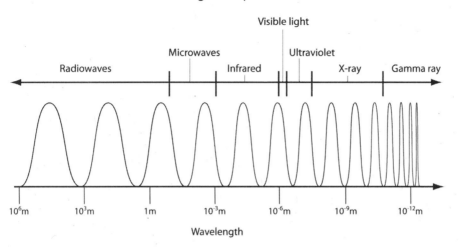

Fig. 2-1:
Electromagnetic Spectrum. The sun produces a wide range of electro-magnetic radiation from gamma rays to radiowaves, shown here. PV cells convert visible light into electric energy, specifically, direct current electricity.

Electromagnetic radiation from the Sun travels virtually unimpeded through space until it encounters the Earth's atmosphere. In the outer portion of the atmosphere (a region known as the stratosphere) ozone molecules (O_3) in the ozone layer absorb much (99%) of the incoming ultraviolet radiation, dramatically reducing our exposure to this potentially harmful form of solar radiation. As sunlight passes through the lower portion of the atmosphere (the troposphere), it encounters clouds, water vapor, and dust. These may absorb the Sun's rays or reflect them back into space, reducing the amount of sunlight striking the Earth's surface, which is referred to as *irradiance*.

Irradiance

Technically speaking, irradiance is the amount of solar radiation striking a square meter of the Earth's atmosphere or the Earth's surface.

Irradiance is measured in watts per square meter (W/m^2). Solar irradiance measured just before the sun's radiation enters the Earth's atmosphere is about 1,366 W/m^2. On a clear day, nearly 30% of the Sun's radiant energy is absorbed or reflected by dust and water vapor in the Earth's atmosphere. (Absorbed sunlight is converted to heat.) By the time the incoming solar radiation reaches a solar array

Measuring Irradiance

Scientists use a device known as a *pyranometer* to measure solar irradiance. Pyranometers measure both direct and diffuse radiation. A shading device placed over a pyranometer allows it to measure only diffuse radiation. Inexpensive pyranometers can be mounted alongside solar arrays to monitor solar input to monitor the performance of a system.

on our roof tops, the 1,366 W/m² measured in outer space has been winnowed down to 1,000 W/m².

Even though the Sun experiences long-term (11-year) cycles, during which its output varies, over the short term, irradiance remains fairly constant. That said, it is important to note that solar irradiance varies during daylight hours at any given site. During the evening, as you'd expect, solar irradiance is zero. As the Sun rises, irradiance increases slowly but surely, peaking around noon. From noon until sunset, irradiance slowly decreases, falling once again to zero at night. These changes in

irradiance are determined by the angle at which the sun's rays strike the Earth, which changes by the minute as the Earth rotates on its axis. The angle at which the sun's rays travel through the atmosphere influences both energy density (Figure 2-2) and the amount of atmosphere through which sunlight must travel to reach the Earth's surface (Figure 2-3).

As shown in Figure 2-2, low-angled sunlight delivers much less energy per square meter than high-angled sunlight, resulting in lower energy density. Early in the morning, then, irradiance is reduced. As the Sun makes

Fig. 2-2: *Energy Density. Surfaces perpendicular to the incoming rays absorb more solar energy than surfaces not perpendicular, as illustrated here.*

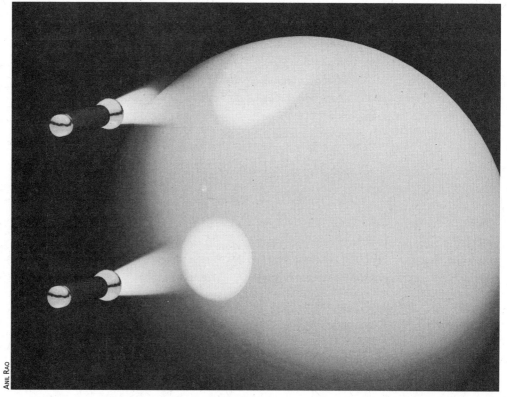

ANIL RAO

its way across the sky, however, energy density and irradiance increase.

Irradiance is also influenced by the amount of atmosphere through which the sunlight passes, as shown in Figure 2-3. The more atmosphere, the more filtering of incoming solar radiation, the less sunlight makes it to Earth, and the lower the irradiance. We'll elaborate on this concept shortly.

Irradiation

Irradiance is an important measurement, but what most solar installers need to know is irradiance over time — the amount of energy they can expect to capture. Irradiance over a period of time is referred to as *solar irradiation*. It's expressed as watts per square meter striking the Earth's surface (or a PV module) for some specified period of time — usually an hour or a day. The units of hourly irradiation are expressed as watt-hours per square meter. For example, 500 watts of solar energy striking a square meter for an hour is referred to as 500 watt-hours per square meter.

To keep irradiance and irradiation straight, you may find it useful to think of irradiance as an instantaneous measure of power (power is measured in watts as noted in Chapter 1).

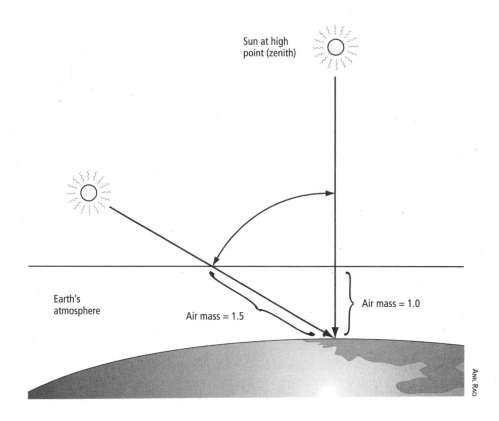

Sun at high point (zenith)

Earth's atmosphere

Air mass = 1.5

Air mass = 1.0

ANIL RAO

Fig. 2-3:
Atmospheric Air Mass and Irradiance. Early and late in the day, sunlight travels through more air in the Earth's atmosphere, which decreases the amount of energy reaching a solar electric array, decreasing its output. Maximum output occurs when the sun's r ays pass through the least amount of atmosphere, at solar noon. Three-quarters of the daily output from a solar array occurs between 9 AM to 3 PM.

Fig. 2-4: *Solar Irradiance and Solar Irradiation. This graph shows both solar irradiance (watts per square meter striking the Earth's surface) and solar irradiation (watts per square meter per day). Note that irradiation is total solar irradiance over some period of time, usually a day. Solar irradiation is the area under the curve.*

Irradiation, on the other hand, is a measure of power over time and is, therefore, a measure of energy. Teachers help students keep the terms straight by likening irradiance to the speed of a car. Like irradiance, speed is an instantaneous measurement. It simply tells us how fast a car is moving. Irradiation is akin to the distance a vehicle travels. Distance, of course, is determined by multiplying the speed of a vehicle by time it travels at a given speed. In a car, the longer you drive at a given speed, the greater the distance you'll cover. In solar energy, the greater the irradiance, the greater the solar irradiation during a given time interval.

Figure 2-4 illustrates the concepts graphically. As shown, irradiance is the single black line in the graph — the number of watts per square meter at any one time. The area under the curve is solar irradiation, the total solar irradiance during a given period — in this graph,

it's the irradiance during a single day. The accompanying sidebar provides yet another way to think about irradiance and irradiation.

Solar irradiation is useful to professional installers and do-it-yourselfers when sizing PV systems — that is, determining the proper size needed to meet the electrical demands of customers. Solar irradiation is also useful to those utilities that base rebates on the projected electrical production of their customers' PV systems.

Direct vs. Diffuse Radiation

Now that you understand irradiance and irradiation, let's take a closer look at solar radiation (electromagnetic radiation released from the Sun) to see what happens to it as it passes through the atmosphere. As shown in Figure 2-5, some sunlight entering the Earth's atmosphere passes through the sky unimpeded. It reaches the Earth's surface directly — that is, without being blocked by clouds, dust, or water vapor. This is referred to as *direct radiation*. The parallel rays of sunlight in direct radiation produce distinct shadows. The remainder of the sun's radiation is either absorbed or scattered by clouds, water molecules, and dust suspended in the atmosphere. Solar radiation absorbed by components of the Earth's atmosphere is converted to heat. Solar radiation reflected off is dispersed in many directions and is known as diffuse radiation. As illustrated in Figure 2-5, some of this diffuse radiation is reflected into outer space; the rest radiates down to Earth. Unlike direct radiation, diffuse radiation produces no shadows.

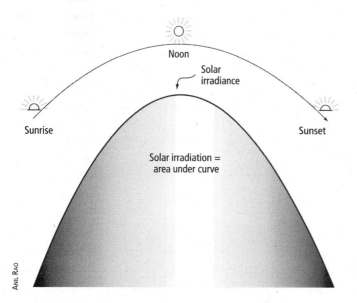

ANIL RAO

On a clear day, direct radiation accounts for 80 to 90% of the sunlight striking the Earth's surface. Diffuse radiation accounts for the rest. On a cloudy day, 100% of the sunlight striking the Earth's surface is diffuse radiation.

The relative proportion of direct and diffuse radiation striking the Earth's surface is influenced by a number of factors. One of the most important is the amount of atmosphere through which solar radiation must pass to reach the Earth's surface. The amount of atmosphere, in turn, affects the amount of

Irradiance vs. Irradiation

As you may recall from Chapter 1, *watts* is a measure of power and *watt-hours* is a measure of energy. *Irradiance* is therefore a measure of the *power density* of electromagnetic radiation falling on a surface. As noted in the text, it is measured in watts per square meter or kilowatts per square meter. *Irradiation* is a measurement of the *energy density* of electromagnetic radiation falling on a surface, and is measured in watt-hours per square meter or kilowatt-hours per square meter.

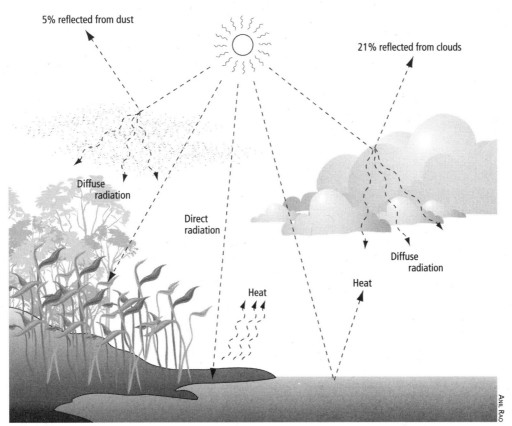

5% reflected from dust

21% reflected from clouds

Diffuse radiation

Direct radiation

Diffuse radiation

Heat

Heat

ANIL RAO

Fig. 2-5:
Diffuse and Direct Radiation. Sunlight streaming through the Earth's atmosphere may pass through unimpeded. This is known as direct radiation. Or, it may be scattered by dust or clouds. Scattered light is known as diffuse radiation.

absorption and scattering that occur. On any given day, for instance, early morning and late day sun rays travel through more atmosphere than when the Sun is higher in the sky (Figure 2-3). The more atmosphere through which sunlight must travel, the more absorption and scattering take place by clouds, water vapor, and dust in the atmosphere. The more absorption and scattering that occur, the less direct radiation that strikes the Earth's surface. The lower the amount of direct radiation, the lower the irradiance and the lower a PV array's electrical output.

Time of day is not the only factor that affects irradiance and irradiation. Elevation also plays an important role. All things being equal, high-elevation regions receive more direct sunlight than low-elevation regions because there's less atmosphere to absorb and scatter incoming solar radiation. (Of course, the amount of direct radiation striking the Earth also depends on the time of year; we'll discuss in the next section).

The vast majority of solar modules absorb both direct and diffuse radiation and convert it into electricity. However, only direct radiation is captured by *concentrating solar collectors*. Concentrating solar collectors are equipped with concentrating lenses or mirrors that focus solar energy onto the PV cells, greatly

Fig. 2-6: Concentrating Solar Cells. Some concentrating solar systems like the one shown here use inexpensive plastic lenses to focus (concentrate) sunlight onto high-efficiency PV cells, minimizing the size of the solar cell required and increasing output. This photo is of the Amonix High Concentration Photovoltaic (HCPV) Solar Power Generator which has a generating capacity 53kW and stands about 50 feet tall and 77 feet wide. This array is designed for utility-scale solar power generation.

increasing their output (Figure 2-6). Although they are more efficient than standard flat plate modules, they're not very common in residential and business applications. Efforts are, however, underway to use concentrating collectors in larger commercial arrays.

Peak Sun, Peak Sun Hours, and Insolation

Another measurement installers use when designing PV systems is *peak sun*. Peak sun is the maximum solar irradiance available at most locations on the Earth's surface on a clear day — 1,000 W/m². One hour of peak sun is known as a *peak sun hour*. Figure 2-6 is a map that shows the daily average peak sun hours in the United States and Canada measured as kW/m²/day.

Peak sun hours is a measure of solar irradiation (watts per m² per day) and is also referred to as solar *insolation* or simply insolation. Insolation is used to compare solar resources of different regions. The annual average insolation in southern New Mexico and Arizona, for instance, is 6.5 to 7 hours of peak sun per day. In Missouri, it's 4 to 4.5 peak sun hours. In cloudy British Columbia, it's about 4.

Installers determine solar insolation by consulting detailed state maps or tables. They use this number to size solar arrays — that is, to determine how large an array must be to meet a customer's needs. As you'll learn in Chapter 4, installers multiply average peak sun hours in a given area by the capacity of the array (in watts) to estimate the average daily output in kilowatt-hours. They then adjust this number to account for efficiency losses and other factors. For example, an installer in New Mexico would calculate the output of an unshaded 1 kW array by multiplying 1 kW by peak sun for the state, which ranges from 6.5 to 7 kilowatt. In this example, the 1 kW array should produce, on average, 6,500 to 7,000 watt-hours (or 6.5 to 7.0 kWh) of electricity per day, minus 30% to compensate for dust and efficiency losses in the system, a topic we'll discuss in Chapter 4. If the array were shaded during part of the day, for example, by a nearby tree, the installer would also need to reduce the estimate accordingly. In a snowy region, an array's output would also need to be adjusted. (Snow that falls on arrays may take a few hours or even a full day to melt off. Until it melts, the snow reduces solar gain by blocking radiation.)

It's important to note that the average peak sun hours per day for a given location doesn't mean that the Sun shines at peak intensity during that entire period. In fact, peak sun conditions — solar irradiance equal to 1,000 W/m² — will very likely only occur one or two hours a day. So how do scientists calculate peak sun hours per day?

Peak sun hours are calculated at a given location by determining the total irradiation received during daylight hours and dividing that number by 1,000 watts per square meter. On a summer day, for instance, the solar irradiance may average 600 watts over 10 hours. As a result, the total solar irradiation is 6,000 watt-hours, and peak sun hours is 6,000 watt-hours divided by 1,000 or 6.0 hours of peak sun. Even though the peak sun hours per day

Fig. 2-7 a and b:
Average Peak Sun Hours per Day These maps shows average peak sun hours per day for the US and Canada. This data is used to estimate the annual electrical output of a solar array.

Peak sun hours per day

■ 1.0 to 1.5
■ 1.5 to 2.0
■ 2.0 to 2.5
▨ 2.5 to 3.0
▨ 3.0 to 3.5
▨ 3.5 to 4.0
▨ 4.0 to 4.5
▨ 4.5 to 5.0
▨ 5.0 to 5.5
▨ 5.5 to 6.0
□ 6.0 to 6.5
□ 6.5 to 7.0
□ 7.0 to 7.5

Kwh/Kw

800- / 900- / 1000- / 1100- / 1200- / 1300- / 1400+
900 / 1000 / 1100 / 1200 / 1300 / 1400

is 6, solar irradiance may have only reached 1,000 W/m² for one to two hours.

The Sun and the Earth: Understanding the Relationships

Now that you understand irradiance, irradiation, peak sun, peak sun hours, and insolation, let's examine the geometric relationships between the Earth and Sun. We'll study the Earth's tilt and the Earth's orbit around the Sun and how they affect the amount of sunlight striking the Earth during the year. In this section, you'll also learn about the altitude angle and azimuth angle, two important factors that affect the amount of sunlight striking

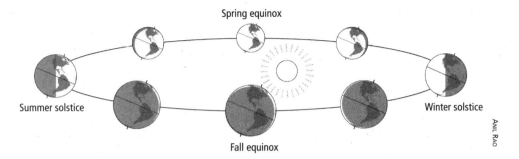

Fig. 2-8: *As any school child can tell you, the Earth orbits around the Sun. Its orbit is not circular, but rather elliptical, as shown here. Notice that the Earth is tilted on its axis and is farthest from the sun during the summer. Because the northern hemisphere is angled toward the sun, however, our summers are warm.*

a solar array. You'll see how these factors affect the orientation and the angle at which the PV solar modules are mounted (this is known as the tilt angle).

Day Length and Altitude Angle: The Earth's Tilt and Orbit Around the Sun

As you learned in grade school, the Earth orbits around the sun, completing its path every 365 days. As shown in Figure 2-8, the Earth's orbit is elliptical, so the distance from the Sun varies. The Earth is closest to the Sun during the winter and farthest from the Sun in the summer. It's obviously not the distance from the Sun, then, that determines our seasons and the amount of sunlight that strikes the Earth. It's the Earth's tilt.

As shown in Figure 2-9, the Earth's axis is tilted 23.5°. The Earth maintains this angle throughout the year as it orbits around the Sun. Look carefully at Figure 2-8 to see that the angle remains fixed — almost as if the Earth were attached to a wire anchored to a fixed point in outer space. Because the Earth's

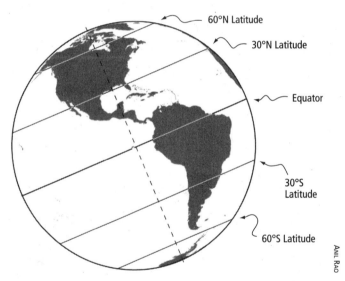

Fig. 2-9: *The Earth is tilted on its axis of rotation, a simple fact with profound implications on solar energy use on Earth, as explained in the text.*

tilt remains constant, the Northern Hemisphere is tilted away from the Sun during the winter. As a result, the Sun's rays enter and pass through Earth's atmosphere at a very low angle. Sunlight penetrating at a low angle passes through more atmosphere and is therefore absorbed or scattered by dust and water vapor, as shown in Figure 2-3. This, in turn, reduces irradiance, reducing the output of a solar array.

Irradiance is also lowered because the density of sunlight striking the Earth's surface is reduced when it strikes a surface at an angle (described earlier). As shown in Figure 2-2, a surface perpendicular to the sun's rays absorbs more solar energy than one that's tilted away from it. (As the sun's rays move away from perpendicular, the surface intercepts less energy.) As a result, low-angled sunlight delivers much less energy per square meter of surface in the winter than it does during summer months.

Solar gain is also reduced in the winter because days are shorter — that is, there are fewer hours of daylight during winter months. Day length is determined by the angle of the Earth in relation to the Sun. During the winter in the Northern Hemisphere, most of the Sun's rays fall on the Southern Hemisphere.

All three factors combine to reduce the amount of solar energy available to a PV system in the winter. They are also responsible for the cooler temperatures of fall, winter, and early spring, all of which increase the efficiency of

Fig. 2-10: *Summer and Winter Solstice. Notice that the Northern Hemisphere is bathed in sunlight during the summer solstice because the Earth is tilted toward the sun. The northern Hemisphere is tilted away from the Sun, during the winter.*

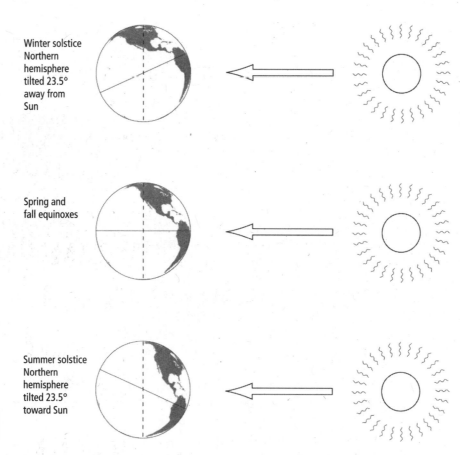

Winter solstice Northern hemisphere tilted 23.5° away from Sun

Spring and fall equinoxes

Summer solstice Northern hemisphere tilted 23.5° toward Sun

ANIL RAO

solar electric modules. This, in turn, slightly offsets reduced solar gain.

In the summer in the Northern Hemisphere the Earth is tilted *toward* the Sun, as shown in Figure 2-8 and 2-10. This results in several key changes. One of them is that the Sun is positioned higher in the sky. As a result, sunlight streaming onto the Northern Hemisphere passes through less atmosphere, which reduces absorption and scattering. This, in turn, increases solar irradiance, which increases the output of a PV array. Because a surface perpendicular to the sun's rays absorbs more solar energy than one that's tilted away from it, the Earth's surface intercepts more energy during the summer as well. Put another way, the high-angled sun delivers much more energy per square meter of surface area than in the winter. Days are also longer in the summer. All these factors increase the output of an array, although the warmer days of summer slightly offset these gains (an array's output decreases as temperatures increase at any given irradiance level).

Figure 2-11 shows the position of the Sun as it "moves" across the sky during different times of the year as a result of the changing relationship between the Earth and the Sun. This illustration shows that, as just discussed, the Sun "carves" a high path across the summer sky. It reaches its highest arc on June 21, the longest day of the year, also known as the summer solstice. Figure 2-11 also shows that the lowest arc occurs on December 21, the shortest day of the year. This is the winter solstice. The angle between the Sun and the horizon at any time during the day is referred to as the *altitude angle*.

As shown in Figure 2-11, the altitude angle decreases from the summer solstice to the winter solstice. After the winter solstice, however, the altitude angle increases, growing a little each day, until the summer solstice returns. Day length changes along with altitude angle, decreasing for six months from the summer solstice to the winter solstice, then increasing until the summer solstice arrives once again.

Fig. 2-11: *Solar Path. This drawing shows the position of the Sun in the sky during the day in the summer and winter solstices and the spring and fall equinoxes. This plot shows the solar window — the area you want to keep unshaded so the Sun is available for generating solar electricity.*

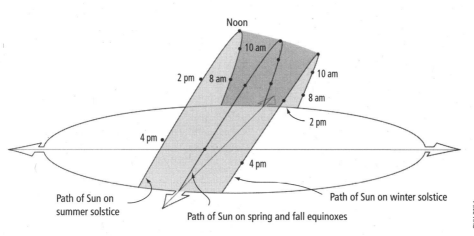

Noon

10 am
2 pm   8 am
10 am
8 am
2 pm
4 pm
4 pm

Path of Sun on summer solstice
Path of Sun on winter solstice
Path of Sun on spring and fall equinoxes

ANIL RAO

The midpoints in the six-month cycles between the summer and winter solstices are known as *equinoxes*. The word *equinox* is derived from the Latin words *aequus* (equal) and *nox* (night). On the equinoxes, the hours of daylight are nearly equal to the hours of darkness. The spring equinox occurs around March 20 and the fall equinox occurs around September 22. These dates mark the beginning of summer and fall, respectively.

The altitude angle of the Sun at any given time during the day is also determined by the rotation of the Earth on its axis. As seen in Figure 2-11, the altitude angle increases between sunrise and noon, then decreases to zero once again at sunset. In addition to the change in the altitude angle of the Sun, the Sun's position in the sky relative to a fixed point, such as a PV array, also changes by the minute. Scientists locate the Sun's position in the sky in relation to a fixed point by the *azimuth angle*. As illustrated in Figure 2-12, true south is assigned a value of 0°. East is +90° and west is -90°. North is 180°. The angle between the Sun and 0° south (the reference point) is known as the *solar azimuth angle*. If the Sun is east of south, the

azimuth angle falls in the range of 0 to + 180°; if it is west of south, it falls between 0 and - 180°. Like altitude angle, azimuth angle changes as a result of the Earth's rotation on its axis.

Implications of Sun-Earth Relationships on Solar Installations

Solar modules can be mounted on four types of racks: fixed, seasonally adjustable fixed, single-axis tracker, or a dual-axis tracker.

Fixed racks are oriented to the south and are set at a fixed tilt angle, usually equal to the latitude of the site. As shown in Figure 2-13, the tilt angle is the vertical angle between the surface of the array and an imaginary horizontal line extending back from the array. Fixed racks are mounted on the roof of buildings or on the ground or even on poles.

Seasonally adjustable fixed racks resemble fixed racks, but can be adjusted to increase or decrease the tilt angle during various seasons. Tilt angle may be increased in the winter to capture more energy from the low-angled winter sun and decreased in the summer to capture more energy from the high-angled summer sun.

Fig. 2-12: Altitude and Azimuth Angles. The altitude angle is the angle of the sun from the horizon. It changes minute by minute as the Earth rotates. It also changes by day. The azimuth angle is the angle of the sun from true south.

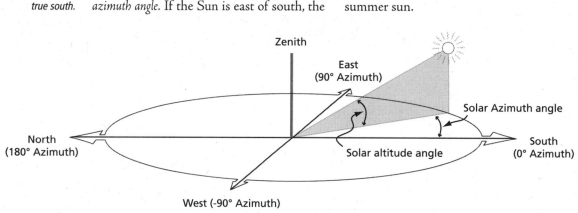

ANIL RAO

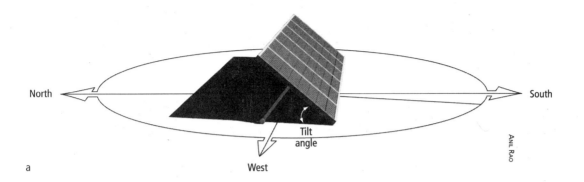

North

South

Tilt
angle

West

ANIL RAO

a

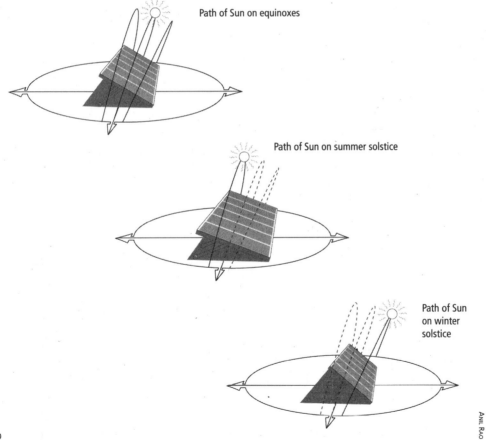

Path of Sun on equinoxes

Path of Sun on summer solstice

Path of Sun
on winter
solstice

ANIL RAO

b

Fig. 2-13 a and b:
Tilt Angle.
(a) The tilt angle of an array, shown here, is adjusted according to the altitude angle of the Sun. The tilt angle can be set at one angle year round, known as the optimum angle, or
(b) adjusted seasonally or even monthly, although most homeowners find this too much trouble.

Single-axis trackers are designed to follow the Sun from sunrise to sunset by automatically adjusting for changes in the azimuth angle of the sun. Single-axis trackers can increase the output of an array up to 30%.

Dual-axis trackers automatically adjust both the tilt angle and the azimuth angle and can increase the output of an array up to 40%, depending on the location.

For optimum solar gain, solar arrays should be pointed toward true south (in the Northern Hemisphere), not magnetic south. What's the difference?

True north and south are measurements used by surveyors to determine property lines. They are imaginary lines that run parallel to the lines of longitude, which, of course,

run from the North Pole to the South Pole. (True north and south are also known as true geographic north and south.) Magnetic north and south, on the other hand, are determined by the Earth's magnetic field. They are measured by compasses. Unfortunately, magnetic north and south rarely line up with the lines of longitude — that is, they rarely run true north and south. In some areas, magnetic north and south can deviate quite significantly from true north and south. How far magnetic north and south deviate from true north and south is known as the *magnetic declination.*

Figure 2-14 shows the deviation of true north and south — the magnetic declination — from magnetic north and south in North America. You may want to take a moment to

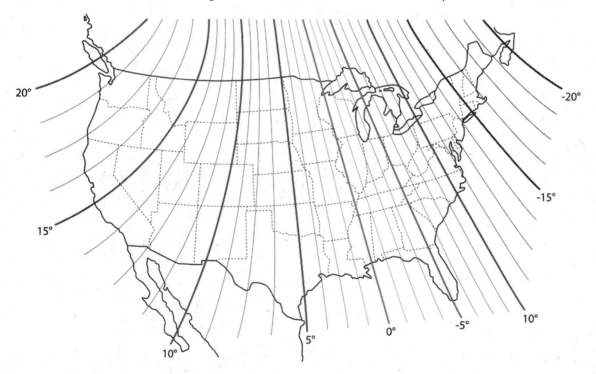

study the map. Start by locating your state and then reading the value of the closest isobar. If you live in the eastern United States, you'll notice that the lines are labeled with a minus sign. This indicates a westerly declination — meaning that true south is located west of magnetic south. If you live in the midwestern and western United States, the lines are positive, which indicates an east declination. That is, true north and south lie east of magnetic north and south.

To determine the magnetic declination of your home or business, you can consult this map, although a local surveyor or nearby airport can provide a more precise reading. Be sure to ask whether the magnetic declination is east or west. Bear in mind that magnetic north and south not only deviate from true north and south, they change very slightly from one year to the next. Surveyors keep track of the annual variation. Also bear in mind that compass readings at any one site may be slightly off. If you take a magnetic reading too close to your vehicle, for instance, the compass may not point precisely to magnetic south. For more on this subject and a failproof way to determine true south, see the accompanying box.

As we will point out in subsequent chapters, orienting a solar array to true south is ideal, but it's not always possible. Don't sweat it. You have some leeway. In fact, orienting an array slightly off true south results in only a very slight decrease in output until the azimuth angle is off by about 25°.

Ideally, a solar array should point directly at the Sun from sunrise to sunset to produce the most energy. Doing so optimizes the energy density falling on the array, which was discussed earlier in the chapter. This is the primary benefit of pointing an array at the Sun. Doing so also ensures sunlight strikes the array head on — that is, perpendicular to the surface of the array — at all times. This minimizes reflection of sunlight off the surface of the module and ensures maximum absorption. (This is a secondary benefit, and not as significant as maximizing energy density.)

The angle at which sunlight energy strikes the surface of an array — or any object, for that matter — is known as the *angle of incidence*. (Technically, the angle of incidence is the angle between incoming solar radiation and a line perpendicular to the surface of the module.) Reducing the angle of incidence increases the irradiance and reduces amount of sunlight that reflects off the array, increasing output. Ideally, the angle of incidence (angle of the incoming solar radiation) should be 0°.

Aligning the array at all times with the Sun to reduce the angle of incidence is possible through the use of trackers, which are actively or passively controlled solar mounting systems. As just noted, single-axis pole-mounted trackers automatically adjust for changes in the azimuth angle. That is, they track the Sun from east to west as it moves across the sky. They do not adjust the tilt angle as the altitude angle of the Sun changes during the day. Dual-axis trackers adjust the altitude *and* azimuth angles. That is, they track the Sun from sunrise to sunset, but also adjust the tilt angle of the array to accommodate changes in the altitude angle.

Finding True South

Most solar installers locate true south by first find-ing magnetic south, using a compass, and then adjusting for magnetic declination, as explained in the text. Unfortunately, this procedure may not always be accurate. One reason for this is that mag-netic fields can be influenced by nearby iron ore deposits. Compass readings are influenced by iron, steel or nearby magnets. A nearby vehicle, keys in your pocket, metal tools, steel posts, cell phones, and audio speakers (which contain magnets) can all alter a compass reading. We call this magnetic devi-ation. To avoid magnetic deviation, be sure to move away from vehicles or metal objects.

Another — failsafe — method of finding true north and south is to head out on a cloudless night and locate the North Star, Polaris. "The North Star is always either exactly or just a few tenths of a degree from true north," say Grey Chisholm, author of "Finding True South the Easy Way" in *Home Power* (Issue 120) notes. How do you locate it?

The North Star is situated at the end of the han-dle of the Little Dipper, as shown in Figure 2-15. It can also be located "by following the two pointer stars on the outside of the Big Dipper's cup," says Chisholm. A line drawn from them will take you directly to the North Star.

To determine true north and south from the North Star, drive a tall fence post into the ground, then move south a bit so you are in line with the post and the North Star. Drive a second fence post into the ground at this point. You've now created a line that runs true north and south. Use this line to orient your PV array.

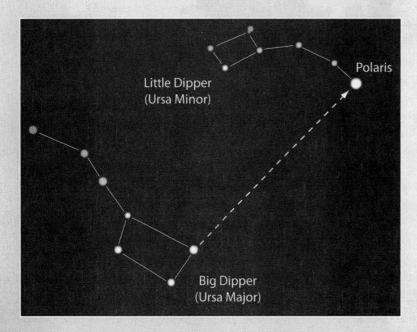

Fig. 2-15: *The North Star. Use the North Star (Polaris) to locate true north.*

Although tracking helps improve the output of an array, most PV modules are mounted on fixed racks — set at a specific angle all year round for convenience. Fixed racks can be attached to poles or mounted on the ground or on the roofs of buildings. When installers mount a fixed rack, they adjust the tilt angle of the array to achieve optimum performance.

As you'll see in Chapter 8, most installers mount arrays at a tilt angle that corresponds to the latitude of the site. In Denver, Colorado, at 40° north latitude, installers typically mount arrays at a 40° tilt angle. In northern Florida, at 30° north latitude, they set the tilt angle of arrays at 30°.

Orienting a solar array as close to true south as possible with a tilt angle that matches the local latitude generally provides the best year-round performance for a fixed array. However, to optimize performance in the summer, the tilt angle should be reduced by about 15°, and to increase output in the winter, the tilt angle should be increased by about 15°.

As you'll learn in Chapter 8, solar sites vary considerably. Most sites experience more sunlight during the summer than winter and some locations receive virtually all of their sunlight during the summer. In Pemberton, British Columbia (near Whistler), at 50° north latitude, for instance, winters are extremely cloudy (averaging only 0.8 to 2 peak sun hours per day), as shown in Table 2-1. Summers are much sunnier with average peak sun hours per day ranging from 3.6 to 5.6. Based on the rule of thumb, the tilt angle of the array should be set at 50°. However, because summer's the sunnier season, reducing the tilt angle of a fixed array allows a homeowner to capture the more-abundant high-angled late spring, summer and early fall sun, improving annual performance of the PV system.

Conclusion

Solar energy is an enormous resource that could contribute mightily in years to come. It could heat and cool our homes and power lights and dozens of household electronic devices from dishwashers to television sets. It could even provide power to electric cars used by commuters. To make the most of it, though, a solar system has to be installed properly — that is, oriented properly to take into account the Sun's daily and seasonally changing position in the sky. In the next chapter, we turn our attention to the technologies that capture solar energy, solar cells.

Table 2.1 Solar Insolation in Pemberton, B.C. Monthly Averaged Insolation Incident On A Horizontal Surface (kWh/m²/day)													
Lat 50.367 Lon -122.85	Jan	Feb	Mar	Apr	May	Jun	Jul	Aug	Sep	Oct	Nov	Dec	Annual Average
10-year Average	1.02	1.77	2.90	4.04	4.92	5.28	5.59	4.82	3.64	2.18	1.14	0.81	3.18

UNDERSTANDING
SOLAR ELECTRICITY

Many people view solar electricity as a new and novel technology. Or, if they're familiar with the space program, they may view solar electricity as a by-product of the aerospace industry's efforts to power satellites, an effort that dates back to the early 1950s. Fact is, the development of solar electricity began a very long time ago — well over 100 years ago.

In this chapter, we'll briefly review the history of solar electricity, then peek inside solar cells to see how PV modules convert sunlight energy to electricity. We'll discuss the types of modules on the market today and some of the newest and most promising PV technologies working their way to market. In this chapter, we'll also introduce you to some terminology that will be helpful to you, no matter whether you are a homeowner or business owner who wants to install a PV system or an individual looking to start a career in solar electricity.

Brief History of PVs

Like so many ideas whose time has come, solar electricity got its start a long time ago in a series of puzzling and seemingly unimportant discoveries. Edmund Becquerel, a 19-year-old French scientist, began the journey of discovery in 1839 when he immersed two brass plates in a conductive liquid and shined a light on the apparatus. He found that an electrical current was produced.

Because electricity had few uses at the time, his discovery sparked little interest. Becquerel, who'd go on to contribute mightily to physics, had apparently uncovered another scientific oddity of little value.

Then, in 1873, a British engineer, Willoughby Smith, made another important discovery while working on undersea telegraph cables. Smith used a device to test the cables as they were being laid down. This test device used bars of selenium. Much to his surprise, he found that the resistance to the flow of electricity in the selenium varied in the lab. Sometimes the bars exhibited high resistance, sometimes they exhibited low resistance. Upon further investigation, Smith discovered that resistance changed when light was shone on the bars. Moreover, the change in resistance was proportional to the amount of light. The more light was shined on the bars, the lower the resistance. Although Smith could not explain his discovery, today we realize that light aided the flow of electrons through the selenium bars. This puzzling phenomenon was later confirmed by researchers W.G. Adams and R.E. Day. They found that light shining on selenium created a tiny electrical current.

Smith's discovery led to applied research designed to find ways this phenomenon could be put to use. The first breakthrough came a few years later, when an American inventor, Charles Fritts, devised the very first electricity-producing solar cell. Fritts' solar cell consisted of a thin layer of selenium to which he applied an ultrathin transparent gold film. When exposed to light — even dim or diffuse light — this device produced a tiny electrical current. Although Fritts could reliably repeat his experiment, engineers were skeptical. How, they wondered, could this device produce electric energy without burning some type of fuel? Some physicists even claimed that Fritts' solar cell violated the laws of physics, specifically the laws of thermodynamics. The first law of thermodynamics states that energy can neither be created nor destroyed. With no apparent source of energy, many engineers viewed Fritts' device with great skepticism.

Undaunted, Fritts sent his solar cells to Werner Siemens, a prominent German inventor and industrialist. Siemens tested the device and confirmed that it did indeed produce electricity when illuminated by sunlight. How it worked, Siemens could not say. But it did work.

As with many other discoveries of the time, it took scientists several decades to understand the inner workings of Fritts' simple solar cell. Today, thanks to the introduction of quantum mechanics, scientists realize that light striking selenium atoms energizes their outer shell electrons. These energized electrons can be forced to flow through an external circuit. This effect is known as the *photovoltaic effect*. It's the basis of all solar cells.

Although Fritts' solar cells worked, they were extremely inefficient at converting sunlight to electricity. The scientist could never achieve efficiencies greater than 1%. Because of the low efficiency and the high cost of selenium, most scientists dismissed this technology as impractical. One exception, however, was the visionary German scientist Bruno Lange, who was also working on solar cells similar in design to Fritts'. In 1931, Lange prophesized that "In the not distant future, huge plants will employ thousands of these plates to transform sunlight into electric power ... that can compete with hydroelectric and steam-driven generators in running factories and lighting homes." Unfortunately, Lange's solar cells were also terribly inefficient.

High costs and low efficiency impaired the development of solar electricity for many years, but in the early 1950s, research in solar electricity took an unexpected turn — a turn that would have profound consequences for this floundering technology. Two researchers, Calvin Fuller and Gordon Pearson, were experimenting with silicon rectifiers at Bell Labs. (A rectifier is a device that converts AC to DC electricity.) Fuller and Pearson discovered that the efficiency of the rectifiers varied based on the amount of light shining on them. Their studies also showed that sunlight striking their silicon-based rectifier caused electrons to flow through them. Quite by accident, then, they had discovered another solar cell — this one, a silicon-based device that would revolutionize the industry. What is more, the efficiency of their solar cells was four times greater (4%) than that achieved by their predecessors, the selenium solar cells.[1] Additional research and development work performed by Fuller and Pearson with Daryl Chapin (who'd worked on selenium cells at Bell Laboratories) boosted the efficiency even more — to 6%.

Excited by the prospects of their researchers' findings, Bell Laboratories began to explore ways to commercialize silicon solar cells. One application that the company hoped would prove lucrative was the use of PV cells to provide electricity to amplify long distance phone signals at remote repeater stations. (A repeater station is an automated facility that greatly extends the range of communications by amplifying telephone signals. It consists of a receiver tuned to one frequency and a transmitter tuned to another, linked by a controller. When the receiver receives a weak signal, the controller activates the transmitter, which retransmits the signal at a higher strength.

Repeaters are usually installed on top of tall buildings or on mountains.) Other companies pursued the use of PVs as well, for example, to power remote buoys for the Coast Guard and lookout towers for the US Forest Service. Unfortunately, the cost of the early PVs were too high and the technology seemed doomed.

But then came satellites. To power satellites, the aerospace industry had concluded that it either had to install batteries or find a way to create energy in space. The energy sources, however, needed to be lightweight, compact and reliable. Batteries failed on all three counts. Not only were they heavy and rather bulky, batteries had a limited lifespan. Solar cells, on the other hand, were relatively lightweight and could generate a continuous supply of electrical energy from the intense solar energy found in outer space — and they could do it for many years.

Since 1958, solar cells have powered virtually all satellites launched into space. Although early solar cells were extremely expensive, that didn't matter to the National Aeronautics and Space Administration (NASA) or the Department of Defense (DOD). There were no other safe, practical options.

Although the space industry owes a deep gratitude to solar cells, PVs today also owe their existence to the space program. That said, we should point out that the solar cells on your neighbor's roof are not the same as those used to power satellites. Nevertheless, these closely related cousins could become an important source of electricity in the not-too-distant future, making the German solar scientist Bruno Lange's prediction come true.

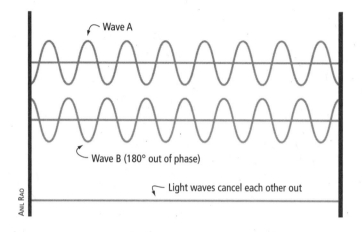

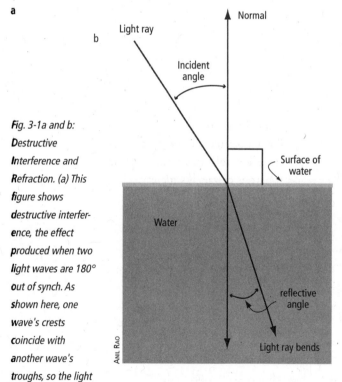

ANIL RAO

Fig. 3-1a and b: **Destructive Interference and Refraction.** (a) This figure shows destructive interference, the effect produced when two light waves are 180° out of synch. As shown here, one wave's crests coincide with another wave's troughs, so the light waves cancel each other out. (b) Refraction is the bending of light as it passes from one transparent medium to another, for example, air to water or air to glass. Both destructive interference and refraction are characteristics of wave behavior.

What is a PV Cell?

What exactly is a solar cell and how does it operate?

Photovoltaic cells are solid-state electronic devices like transistors, diodes and other components of modern electronic equipment. All these devices are referred to as solid-state because electrons flow through solid material within them. They replaced vacuum tubes in which electrons flowed through a vacuum inside a glass tube.

Most solar cells in use today are made from one of the most abundant materials on the planet, silicon. To understand how a PV cell functions, though, you must understand a bit about the atomic structure and the behavior of electrons in silicon atoms. Before we explore these topics, let's take a deeper look at light — the source of energy.

The Nature of Light

For years, scientists thought that solar radiation was emitted and transmitted as waves. However, studies have shown that light also exhibits properties of particles. Today, physicists assert that light has a mysterious dualistic nature. That is, it behaves like waves *and* particles. This phenomenon, which has puzzled many a high school physics student over the years, is referred to as the *wave-particle duality*.

Light exhibits wave behavior when light waves interact with one another. During such times, light waves can cancel each other out. This phenomenon, known as *destructive interference*, is illustrated in Figure 3-1a. Light also exhibits wavelike behavior when it interacts with matter, for example, as light passes from air into a lens it changes speed and bends. This phenomenon is known as refraction and is illustrated in Figure 3-1b. However, light can also exhibit particle behavior when interacting with matter. For example, when light strikes solid materials it transfers small amounts of energy to the atoms in these materials. It's this energy that PV cells use to generate electricity.

Light striking PV cells is transmitted through space in discrete parcels, or packets of energy, called *photons*. A photon is an elementary particle, a carrier of all forms of electromagnetic radiation. A photon differs from many other elementary particles, such as electrons, because it has zero rest mass. As a result, it can travel through a vacuum at the speed of light.

Conductors, Semiconductors and Insulators

To understand how photons of light are converted to electricity in PV cells, let's begin by reviewing atomic structure. As most readers know, an atom is made up of a nucleus, an extremely dense region containing positively charged protons and uncharged neutrons. Orbiting around the nucleus are extremely low mass, negatively charged particles known as electrons. They occupy a region called the electron cloud. It constitutes the bulk of the atom's volume. Electrons live in different regions of the electron cloud. Some orbit more closely around the nucleus than others.

In electrical terms, elements can be classified in three broad categories: conductors, insulators and semiconductors. What's the

Understanding Silicon

Silicon PV cells are made from the element silicon. Silicon is refined from quartz, which is made of silicon dioxide, and from the type of sand that contains quartz particles.

difference? In other words, what makes one atom such as silicon a semiconductor while another like copper a conductor while yet another is an insulator?

The answer is the number of electrons in the outermost region of the electron cloud of the atoms. These electrons are called *valence electrons*. The region of the electron cloud they occupy is known as the *valence shell*.

The valence electrons are responsible for the chemical properties of atoms — that is, they determine how atoms bond to one another to form molecules. The valence electrons also play a key role in determining the electrical properties of various materials. To understand what this means, let's start with conductors.

Conductors are atoms that have only one or two electrons in their outermost shell. These electrons can be easily ejected — that is, forced out of their orbit around the nuclei of their atoms — or shared with neighboring atoms. When propelled out of the valence shell, these electrons are said to be pushed into the conduction band. The conduction band is a range of electron energy that's higher than the valence band. Electrons possessing this energy are free to move about and, more important, to accelerate under the influence of an applied electric field. This, in turn, creates an electric current. As a result, excited electrons can flow from one atom to the next, creating an electrical current in a conductor such as a wire made of copper or aluminum.

Insulators, on the other hand, are atoms that have six to seven electrons in their outer valence shells. For reasons beyond the scope of this book, these electrons cannot be easily jolted out of the atom and made to flow. Thus, materials made from such atoms resist the flow of electricity.

Semiconductors occupy a middle ground. They have three to five electrons in their outermost shells. They are, therefore, marginally conductive. Under certain circumstances, these electrons can be "persuaded" to leave their orbit, to enter the conduction band, where they can flow from one atom to another, creating an electrical current.

Silicon atoms, the main component of computer chips, transistors, diodes, and, ah yes, PV cells, are semiconductors. They have four valence shell electrons. These electrons can be jolted into the conduction band to create an electrical current. What boosts them out of their valence shells?

Light.

Photons with sufficient energy can eject electrons from the valence shell, forcing them into the conduction band. When an electron leaves a silicon atom, it creates an empty space known as a *hole* in the atom from which it came. This hole can be filled by an electron ejected from a neighboring atom. Sunlight striking silicon therefore creates a wild session of musical chairs, although no chairs are withdrawn. Holes are continuously created and filled.

This wild movement of electrons does not result in an orderly flow of electrical current. In fact, the electrons knocked loose by photons drift around aimlessly until they find a hole into which they can fall. When that happens, the energy of the photons is converted to heat, not useful electricity.

What solar cells need to produce electricity is a way to prevent electrons from filling empty holes, so they can be collected and sent to an external circuit where they can do useful work. To do that you've got to get creative.

Scientists have found that they can turn this pointless game of musical chairs into electrical current by adding boron and phosphorus atoms to silicon. This process, called *doping*, cleverly forces electrons to flow in only one direction, creating an orderly flow of electrons out of the solar cell. How is this achieved?

How Solar Cells Work

Most solar cells in use today are thin wafers of silicon about 1/100th of an inch thick (they range from 180 µm to 350 µm in thickness). As shown in Figure 3-2, most solar cells consist of two layers — a very thin upper layer and a much thicker lower layer. The upper layer is doped with phosphorus atoms; the bottom layer is doped with boron atoms.

As shown in Figure 3-3, silicon atoms in solar cells have four electrons in their outermost shells. Each of these electrons bonds with a neighboring silicon atom to form silicon crystals.

As shown in Figure 3-3, the upper layer of a PV, known as the *n-layer*, contains phosphorus atoms. Each phosphorus atom has five electrons in its valence shell. The presence of phosphorous atoms introduces surplus electrons in the n-layer. These rogue electrons can be ejected into the conduction band when struck by photons containing a sufficient amount of energy. Here they are free to move about.

As shown in Figure 3-2, the lower layer of the PV cell is known as the *p-layer*. It is doped with boron atoms. They contain three electrons in their outermost (valence) shells. This, in turn, creates "holes" in the crystalline structure of the p-layer. These holes can be filled by free electrons — electrons ejected from neighboring atoms. Because holes created in one atom can be filled with electrons from another atom, the holes appear to move about.

So how does a PV cell work? The simple answer is that sunlight energy causes electrons to be released from silicon atoms in solar cells; these "loose" electrons are gathered by the metal contacts on the front of solar cells, and thus form an electrical current. It is drawn off the solar modules making up a solar array. (Metal contacts draw electrons from each solar cell.) The slightly more detailed story is as follows: When the n-layer of a PV cell is formed, an electric field is created instantaneously in the cell. This electrical field is created by the flow of electrons from the n-layer to the p-layer at the junction of the two. This creates a charge imbalance at the junction. The top part of the junction is positively

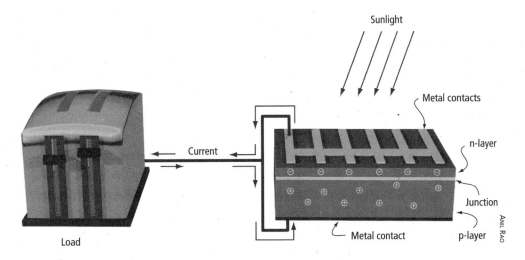

Fig. 3-2: *Cross Section through a Solar Cell. Solar cells like the one shown here consist of two layers of photosensitive silicon, a thin top layer, the n-layer, and a thicker bottom layer, the p-layer. Sunlight causes electrons to flow from the cell through metallic contacts on the surface of most solar cells, creating DC electricity. Solar-energized electrons then flow to loads where the solar energy they carry is used to power the loads. De-energized electrons then flow back to the solar cell.*

Fig. 3-3: *Atomic structure of N- and P-type Layers. This drawing shows the crystalline structure of both the N-type and P-type layers of a solar cell. As shown here, pure crystalline silicon atoms have four electrons in their outermost shell. Each of these electrons bonds with a neighboring silicon atom forming silicon crystals. Because the four electrons bond to neighboring atoms, they are not free to move. Phosphorus atoms in the upper layer (the n-layer) have five electrons in their outermost shell. The surplus electrons in phosphorus atoms in this layer are free to move about when struck by photons containing enough energy to eject the electron. Boron is used to dope the bottom layer. Boron contains three electrons in its outermost shell and thus create holes in the crystalline structure of the boron-doped p-layer that can be filled by free electrons — that is, electrons ejected from neighboring atoms.*

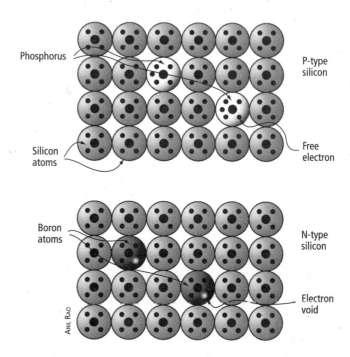

charged. The lower part is negatively charged. This electrical field forces electrons liberated in the p-layer, into the n-layer when sun shines on a PV cell. The electrical field also stops electrons liberated in the n-layer from flowing into the p-layer when sun is shining on a PV cell. (The negatively charged lower part of the junction repels electrons.) The net effect is that electrons can only move in one direction — from the p- and n-layers to the surface of the cell where the contacts are located.

As noted in Chapter 1, numerous solar cells are wired in series in a solar module. Because of this, electrons extracted from one cell flow to the next cell, and then to the next cell, etc., until they reach the negative terminal of the module. After leaving the PV module, electrons flow through an external circuit to a load (a load is any device that consumes electricity). After delivering the energy they gained from sunlight to the load, the electrons return to the positive terminal of the module. They then flow back into the solar cells, filling the holes in the p-layer, permitting the circuit to continue ad infinitum. For a more detailed description, see the sidebar, "How PVs Work."

Types of PV

Solar cells can be made from a variety of semiconductor materials. By far the most common is silicon. Silicon is produced from silicon dioxide, which is derived from two sources: quartzite and silica sand. Quartzite is a rock made entirely of the mineral quartz, made of nearly pure silica (silicon dioxide). Silica sand is relatively pure sand containing a high percentage of silica (silicon dioxide). Geologically, silica sand is derived from quartz. Although not the most efficient in converting sunlight to electricity, silicon dominates the semiconductor market because silicon semiconductors produce the most electricity at the lowest cost.

Three forms of silicon are used to make solar modules: monocrystalline, polycrystalline and amorphous.

Monocrystalline PVs

Monocrystalline cells — a.k.a. single crystal cells — were the very first commercially

How PVs Work

PV cells are relatively simple devices governed by complex physical and chemical interactions. To understand how a PV cell works, we begin with the formation of the top layer, known as the n-layer. Immediately after the n-layer is formed, some of the excess electrons from the phosphorus atoms in this layer diffuse across the junction between the n-layer and the underlying p-layer. These electrons fill the holes in the neighboring p-layer that are created by the addition of boron atoms during the production of a PV cell. This flow of electrons from the n-layer to the p-layer results in both a depletion of electrons and an increase of electron holes in the n-layer near the junction. The flow of electrons into the p-layer near the junction results in a surplus of electrons on that side of the junction and shortage of holes.

The flow of electrons into the p-layer not only results in an excess of electrons on the p-side, it creates a shortage of electrons in the n-layer near the junction. (The p-layer near the junction is negative and the n-layer is positive.) The separation of charge creates a strong electric field across the junction.

When a photon of light ejects an electron into the conduction band in the p-layer and that electron drifts near the junction, it is attracted by the electric field and swept across the junction to the n-layer. Because there are more electrons on the p-side than the n-side, the electron cannot travel in reverse. The negatively charged p-layer surface at the junction blocks the electron from moving back to the p-layer. The same applies to electrons liberated in the n-layer. These electrons cannot flow into the p-layer. They are blocked by the junction (physicists refer to this type of junction as a *junction diode.* Junction diodes allow current to flow in only one direction — in this case, from the p-layer to the n-layer.) Electrons that accumulate in the n-layer are collected at the surface of the solar cell and carried away by a conducting grid on the front surface of the cell.

As electrons move across the junction, from the deeper parts of the p-layer, electron holes are created in the p-layer. These are filled by electrons flowing into the back of the PV cell. A conducting layer is also attached to the back surface of the cell (the p-layer). When light shines on a PV cell, then, electrons flow out of the cell through the front contacts, through a circuit where they can do useful work, and then return via the back terminal. The electrons then fill the holes in the p-layer, ensuring a continuous supply of electricity.

Fig. 3-4:
Monocrystalline
Silicon Ingot. This
ingot is a huge
crystal of silicon that
is sliced to make
monocrystalline
PV cells.

SOLARWOFLD

manufactured solar cells. They are made from wafers sliced from a single crystal of silicon (Figure 3-4). Single crystal ingots are made by melting highly purified chunks of polysilicon (made from sand or quartzite) and a trace amount of boron. Once melted, a seed crystal is dipped in the molten mass (also known as the melt) of silicon and boron. The seed crystal is rotated and slowly withdrawn. As it's withdrawn, silicon atoms from the melt attach to the seed crystal, exactly duplicating its crystal structure. Over time, the crystal grows larger and larger. Eventually, the ingot may grow to a length of 40 inches and a diameter of 8 inches.

Once extracted from the melt, the ingot is cooled. The rounded edges are trimmed and the square or rectangular ingot is then sliced

with a diamond wire saw to produce ultra-thin wafers used to make solar cells. Waste from this process is remelted and reused.

Monocrystalline cells boast the highest efficiency of all conventional PV cells — around 15% — although efficiencies vary from one manufacturer to the next, ranging from 14 to 17%.[2]

Polycrystalline PV Cells

Polycrystalline solar cells are made from silicon with a trace of boron just like monocrystalline cells. To make a polycrystalline cell, however, the molten material is poured into a square or rectangular mold. It is then allowed to cool very slowly. As the ingot cools, many smaller crystals form internally. Once cooled, the cast ingot is removed from the mold, then sliced

Fig. 3-5:
*Polycrystalline
PV Cell. This close-
up photo of a
polycrystalline solar
cell shows the
individual crystals
that form when
molten silicon is
cooled in the mold.*

using a diamond wire saw, creating wafers used to fabricate solar cells.

Polycrystalline solar cells are only about 12% efficient. (Their efficiency varies from one manufacturer to the next, ranging from 11.5 to 14%). Why are polycrystalline cells less efficient than monocrystalline PV cells?

As just noted, polycrystalline PV cells contain numerous small crystals (Figure 3-5). The boundaries between these crystals resist the flow of electrons, reducing the output of a cell. The boundaries also create opportunities for electrons to fill holes in the crystalline structure before the electrons can be whisked away at a PV cell's surface. Both these factors reduce the efficiency of the cell. So why bother making them?

Although polycrystalline cells are less efficient, they require less energy to produce.

Because of this, they're a bit cheaper to manufacture.

Ribbon Polycrystalline PV

While most polycrystalline cells are sliced from ingots, wafers can also be made in long continuous ribbons. Although a dozen different technologies have been developed to make ribbons, only four are currently used. In one of these techniques, shown in Figure 3-6, a seed crystal is attached to two heat-resistant wires called *filaments*. The filaments are immersed in a molten mass of silicon and then slowly drawn from the melt. The ribbon grows linearly as the molten silicon that spans the wires solidifies. Once the ribbon is completely formed, it can be sliced into rectangular wafers. Figure 3-7 shows another, similar process used by Evergreen Solar, a US

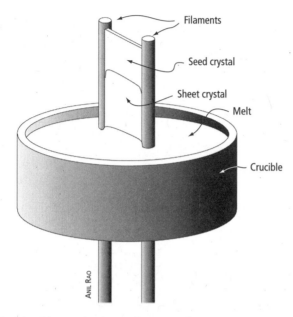

Fig. 3-6 : *Ribbon PV This drawing shows one of several techniques used by PV manufacturers to produce ribbons of PV polycrystalline material. They are sliced to create solar cells mounted in modules.*

company. As you can see, they produce two ribbons simultaneously from each string ribbon furnace.

Although the efficiency of ribbon silicon wafers is slightly lower than other PV cells — it falls in the 11 to 13% range — this technology offers several advantages. One is that it requires considerably less energy than monocrystalline and polycrystalline production. The ribbon technology also eliminates a lot of time-consuming, wasteful and costly ingot slicing.

Adding the N-Layer

All three of the processes described so far result in the formation of the p-layer, the thicker, bottom layer of PV cells. How is the much thinner, phosphorus-containing n-layer added?

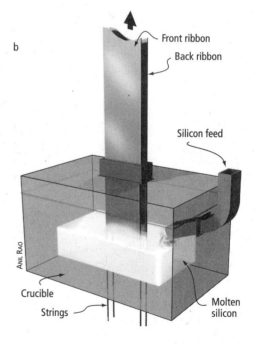

Fig. 3-7: *Ribbon Technology. (a) Photo of solar cells in an Evergreen solar module. (b) This drawing shows the ribbon making technology employed by Evergreen Solar.*

After the p-layer has been created, the wafer is dipped in a solution of sodium hydroxide. This powerful base removes impurities that may have deposited on the wafers during manufacturing. It also etches (roughens) the surfaces of the wafer, which allows subsequent coats to adhere better. Etching also reduces reflection, improving the efficiency of solar cells. (Reduced reflection means more sunlight is absorbed by the cell.)

Once etched, the wafers are placed in a diffusion furnace and heated. Phosphorous gas is introduced into the furnace. In this high-temperature environment, phosphorus atoms penetrate the exposed top surface and sides of the wafer. (The bottom side is masked to protect it from the gas.) This results in the formation of a very thin, but complete n-type layer.

After a wafer is removed from the diffusion furnace, its edges are shaved off — very delicately, of course — to keep the n-layers from contacting the underlying p-layer along the perimeter of the cells.

Metal contacts are then screen-printed onto the face (the n-layer side) of the wafer using a silver paste. Contacts are applied in a grid pattern consisting of two or more wide main strips and numerous ultrafine (hair-thin) strips that attach perpendicularly to the main strips (Figure 3-8). This grid collects the electrons that gather in the n-layer, and forms the negative connection of each cell. As noted earlier, another conducting layer, usually made from aluminum, is applied to the back of each cell.

Because silicon is highly reflective, an antireflective coating of silicon monoxide or other

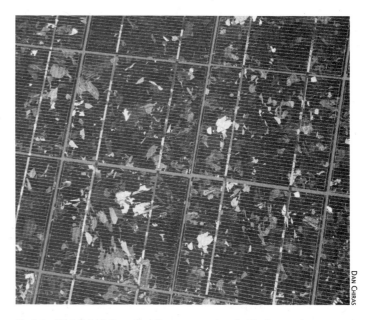

Fig. 3-8: *PV Cell Grid. These fine silver contacts described in the text draw electricity off the surface of the module.*

optical material is applied to the surface of the cells. This coating helps reduce the reflection of light off a module, resulting in greater absorption of the sun's energy and greater output.[3]

Once completed, the cells are individually tested for voltage and current output under controlled conditions using artificial light. The cells are then sorted according to their current output. Modules are assembled using cells with similar output. (In a module, the current is limited to the current generated by the weakest cell.)

Thin-Film Technology

In an effort to produce solar modules at a lower cost — which means using less energy and less material — several manufacturers have turned to a new technology, known as *thin-film*. Unlike manufacturers of previous technologies, thin-film producers manufacture entire modules, rather than individual cells that are later assembled into modules. Skipping the step of assembling cells into modules saves energy, time and money.

In silicon-based thin-film solar designs, known as *amorphous silicon*, silicon is deposited directly onto a metal backing (aluminum, glass, or even plastic), by a technique known as *chemical vapor deposition*. This creates a thin film of photo-reactive material. Once it has solidified, a laser is used to delineate cells and create connections between the newly formed cells.

Energy Payback for PVs

You may have heard people say that it takes more energy to make a PV system than you get out of it over its lifetime. Fortunately, that's not even close to being accurate.

While it takes energy to make solar cells, module- and the remaining components of a PV system, the energy payback is actually amazingly short — only one to two years, according to a study released in 2006 by CrystalClear, a research and development project on advanced industrial crystalline silicon PV technology funded by a consortium of European PV manufacturers (http://www.ipcrystalclear.info/default.aspx). As Justine Sanchez notes in her 2008 article in *Home Power*, "PV Energy Payback," "Given that a PV system will continue to produce electricity for 30 years or more, a PV system's lifetime production will far exceed the energy it took to produce it."

CrystalClear's research showed that it takes two years for a PV system with monocrystalline solar cells to make as much energy as was required to manufacture the entire PV system. The researchers also calculated the energy payback for polycrystalline cells and polycrystalline solar cells manufactured by the ribbon technique. The calculations showed that it took 1.7 years for a polycrystalline system and 1.5 years for modules made from ribbon polycrystalline PVs to produce as much energy as was required to make them. According to the National Renewable Energy Laboratory, thin-film modules, which require even less energy to produce, achieve energy payback in one year.

These studies were performed for sunlight conditions similar to those found in southern Europe with an average insolation of 4.7 peak sun-hours. For those who live in sunnier climates, the energy payback would be even quicker. For those who live in less sunny regions, the payback would be slower.

Most of the energy required to make a PV system goes into producing the modules — about 93% of the entire energy budget is devoted to making modules. As just noted, the most energy-intensive modules are those made from monocrystalline solar cells. Polycrystalline cell modules require 15% less energy to manufacture than monocrystalline modules. Ribbon cell module production requires 25% less energy than monocrystalline and about 12% less than polycrystalline. Thin-film uses even less energy, about 50% less than monocrystalline modules.

Thin-film PVs are currently made from amorphous silicon and three other blends of semiconductor materials. These are cadmium telluride (CdTe), copper indium diselenide (CIS), and copper indium gallium diselenide (CIGS). One of the advantages of amorphous silicon, notes Erika Weliczko, in an article in *Home Power* (Issue 127), is that it can be manufactured in long, continuous rolls, or incorporated onto flexible substrates such as laminates, shingles and roofing — even backpacks.

Thin-film technology has been used for many years to produce miniature solar cells that power calculators and similar devices. Although they perform well, the cells used for these technologies are only about 5% efficient. They also break down in intense sunlight, rendering them useless for solar modules.

To address these issues, thin-film PV manufacturers apply several layers of amorphous silicon — or other materials — to their modules. This increases their efficiency — to around 8 to 9% — and prevents photo degradation.

Efforts are being made to increase the efficiency of this promising technology. By carefully selecting the semiconductors, manufacturers can create solar modules that absorb nearly the entire solar spectrum, making the most of the sun's energy. The result is a more efficient PV cell — one that converts more of the sun's energy into electricity. Researchers at the National Renewable Energy Laboratory in Golden, Colorado, for example, have developed a multilayered thin-film cell. Known as a *multi-junction PV cell*, it consists of numerous thin layers, each containing a different type of semiconductor. Each one absorbs a different wavelength (color) of light. This allows solar cells to absorb more of the sunlight striking them.

Originally developed to power satellites, which require ultra-lightweight components, multi-junction solar cells are generally too costly for business or residential applications, but they are now being used to produce highly efficient concentrated PVs (discussed shortly).

Thin-film technology offers several advantages over single and polycrystalline solar cells. One of the most important is that it uses considerably less energy, by eliminating costly and energy-intensive ingot production and wafer slicing required to make mono- and polycrystalline PV cells. Another huge advantage is that it uses less material. Although silicon is abundant, it is not cheap. In recent years, supplies of this semiconductor have been limited because of competing demand from the computer industry, which uses silicon to make computer chips. Another advantage of thin-film PV is that it is less sensitive to high temperatures. At 100°F, a crystalline module will experience a 6% loss in production while a thin-film amorphous silicon array will experience a 2% loss, making thin-film a good choice for sunny climates, provided you have the roof space. Thin-film is also a bit more shade tolerant than crystalline PV. As Erika Weliczko points out, "many thin-film cells are as long as the module itself, so shading an entire cell is more difficult than traditional 5- or 6-inch square or round crystalline cells."

UniSolar is a company that produces modules that incorporate bypass diodes into each PV cell. These allow current to flow around or bypass shaded cells. (Shading increases resistance in a PV cell, which reduces the flow of electricity through a module.) Yet another advantage of thin-film is that thin-film products are uniformly black, a feature that architects, homeowners and others find appealing.

One of the chief disadvantages of thin-film PVs is their low efficiency. Most thin-film products have efficiencies in the 6 to 8% range. Lower efficiency means larger arrays. In fact, you would need approximately 180 square feet of roof space for a 1 kW PV array made from thin-film vs. 90 square feet for a comparable array made from crystalline PV modules. Another potential downside is that it takes thin-film modules 6 to 12 months to reach their stable, rated output. Crystalline modules stabilize immediately. Because the output of thin-film is 20 to 25% higher at first, care must be taken when sizing conductors and other components like charge controllers and inverters in a PV system so the array doesn't exceed rated capacities. Another key disadvantage of thin-film PVs is that the modules produce high-voltage, low-current electricity. The higher the voltage, the fewer the number of modules that can be placed in series. (The National Electrical Code prohibits voltages in battery-less grid-connected systems to no more than 600 volts.) Inverters for battery-based systems require even lower voltages — 12, 24 and 48 volts. High voltage output, therefore,

makes it difficult to design a system with thin-film modules so that it does not exceed the acceptable input voltage of inverters. (For battery-based systems, installers can use step-down charge controllers, which reduce the voltage of the electricity from the array to the proper battery-charging voltage.)

Rating PV Modules and Sizing PV Systems

PV manufacturers rate modules by various parameters, including rated power, power tolerance, power per square foot, and efficiency. They also include several additional parameters, including short-circuit current, open-circuit voltage, maximum power voltage, maximum power current, maximum system voltage, and series fuse rating. These are used to compare PV modules and to design PV systems to ensure that components can handle the voltage and current produced by the array under all conditions.

Rated power is the wattage a module produces under standard test conditions (STC). This measurement is used to compare one module to another. It is also useful when a professional or do-it-yourselfer is sizing a PV array. As noted in Chapter 1, the standard test conditions used to determine rated power rarely reflect typical operating conditions. A more practical measure is the rated power derived under more typical field conditions, known as PTC. (For more on this, see the sidebar "Rated Power and Capacity" in Chapter 1, p. 4.)

The next measurement, power tolerance, is the range within which a module either overperforms or underperforms its rated power at STC. It may be expressed as a percentage or as watts. Rated power tolerance ranges from -9% to +10%, depending on the module. Some, like Evergreen's new 195, 200, 205 and 210-watt modules have a power tolerance of 0/+5 watts. That means that the 200-watt module, for example, is guaranteed to produce between 200 and 205 watts under STC. By contrast, a 200-watt module with a more typical power tolerance of -5%/+10% could produce anywhere from 190 to 220 watts.

Some manufacturers also list *rated power (watts) per square foot* of module. As the name implies, it is the power output per square foot of module area. This number is useful to those who have a limited amount of space to mount a PV array and want to generate as much electricity as possible in that limited space.

Module efficiency is also a handy number. It's the ratio of output power (watts) from a module to input power (watts) from the sun. A 15% module efficiency means that 15% of the incident (incoming) solar radiation is converted into electricity. Like rated power per square foot, it's handy for those who have limited space.

Several additional parameters, like open circuit voltage and short circuit current, are important when it comes to sizing components of a PV system. To understand them, let's begin with the graph shown Figure 3-9a, known as a current-voltage or I-V curve.

The *current-voltage*, or *I-V*, curve shows the relationship between the current and voltage produced by any PV device — a cell, module or array — under a specific set of conditions (cell temperature and irradiance). The parameters mentioned above represent certain key points along the I-V curve ("I" represents current and "V" represents voltage).

Short-circuit current (Isc) is the current that flows from an array if the output terminals (the positive and negative leads) are connected to one another, that is, shorted. While short-circuiting an electrical device like a battery can be very dangerous, short-circuiting a PV module is not. PV devices are *current limited*, meaning they can only produce a certain amount of current under optimal conditions — and that amount of current will not damage the module.

PV Degradation

Thin-film and crystalline PV modules degrade slowly over time, approximately 0.5 to 1.0% per year. Most modules are warranted by the manufacturer to produce 90% of their minimum peak power in ten years and 80% in 20 to 25 years. Warranted minimum power ratings must be adjusted for rated power tolerance. Most modules have a tolerance of plus or minus 5%, meaning a 100-watt module will produce between 95 and 105 watts under standard test conditions. For the warranty, however, a 100-watt module with a specified power tolerance of plus or minus 5% is assumed to have a minimum peak power value of 95 watts. The module's warranty will be based on this number, not the STC 100 watt rating. In 10 years, then, the module would be warranted to produce 90% of 95 watts under standard test conditions.

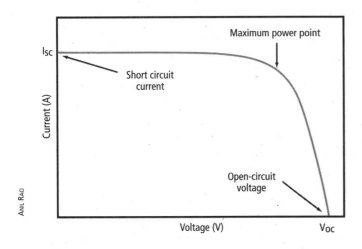

Current (A) ↑

I$_{sc}$

Maximum power point

Short circuit
current

Open-circuit
voltage

Voltage (V) → V$_{oc}$

ANIL RAO

Fig. 3-9: *I-V Curve*

The short-circuit current of an array is used to calculate the amperage rating of wires, fuses and circuit breakers in a PV system because it is the maximum current an array can produce. Appropriate fuses and circuit breakers need to be installed to protect the wires and array from excessive current that could be introduced by a faulty inverter or charge controller, or from other abnormal conditions.

Another measurement that's used to design PV systems is the *open circuit voltage* or Voc. Open circuit voltage, shown in Figure 3-9, is the voltage produced by a PV cell, module or array with no load (inverter or battery) connected. That is, the leads are not connected to a load. As a result, no current can flow.

Open circuit voltage represents the maximum voltage a PV device could produce at standard test conditions. The maximum input voltage rating of inverters and charge controllers used in the system must be higher

than the maximum output voltage of the array to eliminate the possibility of a potentially dangerous over-voltage condition that could cause damage. Because the output of a PV device increases with declining temperature, the maximum output voltage of the array is calculated by multiplying the Voc of the array by an adjustment factor based on the coldest historical temperature for the site.

Figure 3-9 also illustrates how the next three parameters, maximum power voltage, maximum power current, and maximum power point are calculated. As illustrated in the graph, the *maximum power point* is the point on the I-V curve at which the power output (measured in watts) reaches its highest point. Pmp represents the maximum power point. Power is the product of amps x volts. The maximum power voltage (Vmp) and maximum power current (Imp) are those two points on the x and y axes that, when multiplied together, yield the maximum power, or Pmp.

The *maximum power voltage* and *maximum power current* are not particularly important when rating modules and sizing PV systems, but they are extremely important for getting the most power from a PV array. As you'll learn in Chapter 6, most inverters and charge controllers contain circuitry that ensures that the array operates at its maximum power point. This function is known as *maximum power point tracking* (MPPT) and is explained in the sidebar "PV Array Voltage, String Size, and Choosing the Correct Inverter" in Chapter 6, p. 144. In most applications, MPPT will increase the output of a PV array by about

15% per year. The greatest benefits come when you may need it the most — during the winter when the PV cells are cold, the array voltage is high, and the availability of sunlight is lowest.

The *maximum system voltage* is the highest voltage at which the array can safely be operated. On modern modules, this is typically 600 volts. This high maximum system voltage allows installers to connect many modules in series to produce a high-voltage system.

The *series fuse rating* is the value of the fuse that must be connected in series with each module (or series string of modules) to protect the modules from a reverse current that could occur during abnormal conditions.

While all these numbers may seem confusing at first, as you design your system or learn more about PV systems, they'll become much more comprehensible.

Advancements in PV — What's on the Horizon?

Although PVs have come a long way since Becquerel's discovery in 1839, researchers in the private and public sectors continue to advance the technology. They continue to look for ways to improve the efficiency and reduce the cost of solar electric technologies to make solar electricity more affordable to customers and profitable for businesses. This effort has led to some rather promising developments in recent years. Some of these technologies are already available; others are not quite ready for prime time, but could be manufactured commercially in the very near future.

Building Integrated PVs

One of the newest and most popular PV technologies to make its way to the market is *building integrated PV* (BIPV Figure 3-10a and b). BIPV incorporates solar electric generating capacity into the components of a building envelope, for example, the roof, windows, skylights and exterior walls.

On commercial buildings with flat roofs, for instance, you can install a flexible roofing membrane coated with thin-film silicon. For pitched roofs, you can install solar roof tiles — PV modules that replace conventional roof tiles. For standing seam metal roofs, like the one shown in Figure 3-10, you can install a flexible thin-film material known as a PV laminate or PVL. The laminate is encapsulated in UV-stabilized polymers. The product is shipped in rolls. When it is time to apply PVL, a backing sheet is peeled off the laminate, revealing a sticky surface that adheres the product to standing seam metal roofing. PVL could be used for parking structures and the metal roofs of barns, homes, schools, government buildings and so on.

For skylights and windows, you can install glass coated with a thin layer of amorphous silicon. When sunlight shines on the glass, it generates electricity. Some manufacturers sandwich crystalline silicon cells in glass panels that are used to make the roofs of bus stop shelters or canopies. Solar modules can even be mounted in specially designed panels in the exterior walls of buildings — even skyscrapers. Conventional PV modules can be used to make awnings, windows, carports, parking structures and backyard shade structures.

SANYO

a

ENERGY CONVERSION DEVICES/UNI-SOLAR

b

BIPVs are being used primarily in new buildings, although they can be incorporated into existing homes and offices. They offer several advantages over more conventional PV technologies. One of the most important advantages is that they perform multiple functions, which reduces resource depletion and construction costs. For instance, solar awnings, like the ones shown in Figure 3-11, provide shade for windows, lower cooling costs, and eliminate the need for conventional awnings. They do all this while generating electricity.

Another advantage of BIPVs is that they tend to blend in better than conventional PV modules, which many people find appealing. Because of these advantages, BIPV is now one of the fastest-growing parts of the PV industry.

Light-absorbing Dyes

Another new, although not-yet-ready-for-prime-time, PV technology is the *dye-sensitized solar cell* (DSSC). DSSCs contain particles of titanium dioxide (TiO_2) coated by a light-absorbing dye, as shown in Figure 3-12. Sunlight enters the cell through a transparent top layer, striking the dye molecules. Photons with sufficient energy to be absorbed by the dye excite electrons of the dye molecules. These electrons are transferred directly into the conduction band of the titanium dioxide molecules. The electrons then move to the clear surface layer, which doubles as the contact.

Electrons lost from the dye molecules are rapidly replenished by iodide ions (I^-) in the electrolyte surrounding the titanium oxide/dye particles. In the process, iodide ions are converted to trioxide (I^{-3}). Trioxide ions recoup their missing electron by diffusing (moving) to the bottom of the cell. Here they pick up electrons flowing back into the cell via a catalytic conductor through the external circuit — just like an ordinary PV cell.

DSSCs, unlike previous cells described in this chapter, are not solid-state devices. They are photoelectrochemical cells. Although they rely on the photoelectric effect to generate electrons (in the dye) to create current, they also rely on ions in the electrolyte to transfer electrons to the dye molecules.

This technology uses low-cost materials and can be used to manufacture flexible sheets, as in thin-film technology. Moreover, cells can be assembled using much less energy and much simpler and less expensive equipment than is required to manufacture conventional

Fig. 3-11: *PV Awning. Solar awnings, like the one shown here, provides shade for windows and walls, eliminating the need for conventional awnings. Like other forms of building integrated PVs, they perform more than one function. Awnings shade windows and walls and thus cool buildings in the summer and produce electricity year round.*

LIGHTHOUSE SOLAR IN BOULDER, CO

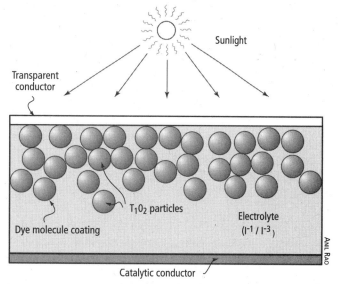

Sunlight

Transparent conductor

T_1O_2 particles

Dye molecule coating

Electrolyte (I^{-1} / I^{-3})

Catalytic conductor

ANIL RAO

Fig. 3-12: *Photoelectrochemical PV Cell. This drawing shows the anatomy of one of the newest solar cells under development.*

monocrystalline and polycrystalline PVs. As a result, they should be cheaper to manufacture.

Although the conversion efficiency is much lower than the best thin-films PVs on the market, lower production costs could enable these cells to compete with electricity generated by coal and other fossil fuels within the next decade or so.

Organic Solar Cells

Another technology that holds promise for the future is the organic or polymer solar cells. Organic solar cells consist of thin plastic films (typically 100 nm) containing semiconductor materials with tongue-twisting names such as polyphenylene vinylene, copper phthalocyanine (which is a blue or green organic pigment), and carbon fullerenes. These molecules are the source of photo-excited electrons.

In the carbon fullerene cells, a polymer (plastic) releases electrons when struck by light. The neighboring carbon fullerene molecules accept the electrons. How the cells generate electricity from this point, however, requires an advanced degree in physics. If you're interested, you can find excellent descriptions on the Internet.

Organic cells are inefficient — so far, the highest efficiencies are around 6.5% — but they could potentially be produced very cheaply, according to some experts. How cheaply? Most conventional solar modules current cost around $5 per watt (of generating capacity). One company now making organic polymer cells believes that they will be able to produce organic polymer solar cells that cost $1 per

watt. Eventually, they predict costs as low as 10 cents per watt.

Bifacial Solar Cells

Yet another promising technology is the bifacial solar cell, also known as a hybrid solar cell. According to Sanyo, manufacturer of a bifacial module, bifacial solar cells consist of a monocrystalline silicon wafer sandwiched between two ultra-thin amorphous silicon layers. Because of this, these modules can capture sunlight falling on both sides of their cells. (This feature is the reason they're called bifacial modules.)

In a bifacial module, the front side of the panel generates electricity from direct and diffuse solar radiation (Figure 3-13). The backside generates electricity from diffuse light from the sky as well as reflected light from surrounding surfaces. Under the right conditions, bifacial modules can produce more power than conventional single-sided modules.

Concentrating PV

The efficiency of solar modules can be increased by using lenses and mirrors to concentrate sunlight on PV cells (Figure 3-14). Concentrating sunlight greatly increases the solar input to the cells. The more sunlight that can be focused on a PV, the more electricity it generates. Note that most concentrating collectors must be mounted on trackers — devices that follow the Sun across the sky — to operate properly.

Concentrating solar collectors are not a new idea. Early attempts, however, proved

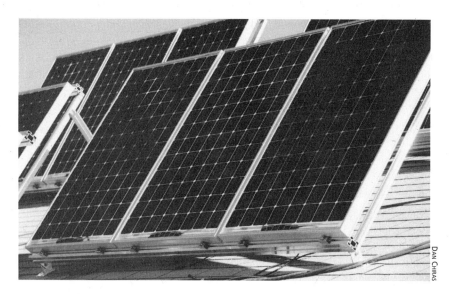

a

Fig. 3-13 a and b: *Bifacial Modules.*
(a) Bifacial modules harvest solar energy off both sides of the panel, although the front side shown in this photograph is by far the most productive source of electricity.
(b) Sunlight reflecting off the light-colored roofing illuminates the array, boosting its output.

disastrous, as concentrated sunlight burned out PV cells. To avoid this problem, PV cells are cooled. Cooling is achieved by use of a heat sink, a device that draws heat off the cells.

Concentrating solar systems can boost PV efficiency to around 45%. Because they operate at a higher efficiency, they require less semiconductor material than other systems. These systems can also use the multi-junction PV cells, which are inherently more efficient.

Concentrating systems do have some limitations. Although they operate very efficiently on clear, sunny days, they function very poorly — or not at all — on cloudy days. That's because clouds scatter incoming solar radiation, increasing the amount of diffuse light. Diffuse light cannot be concentrated like direct light.

b

AMONIX

Fig. 3-14:
*Concentrating Solar
Collectors. Solar
collectors like these
large commercial
arrays use lenses
to concentrate
sunlight on PV cells,
dramatically boost-
ing their output.*

Conclusion: Should You Wait for the Latest, Greatest New Technology?

Knowing that new PV technologies are in the offing, many people ask if they should wait a bit before they invest in a solar electric system. Does it make sense to delay your installation until the new, more efficient PV technologies hit the market?

This is a fair question, but the answer is no.

Newer, more efficient PVs are certainly on their way, however, one of the key considerations when installing a PV system is not the efficiency of the PV modules, but the cost of the modules based on the installed capacity. Those of us in the business refer to this as the *cost per watt of installed capacity*.

What you'll find when comparing modules on this basis is that new technologies, while more efficient, typically cost more per watt of installed capacity. A 2 kW system that

utilizes the most efficient modules on the market, for instance, may cost 10 to 30% more to purchase than a 2 kW system that uses slightly less efficient modules. Yet both arrays produce the same amount of electricity. Why pay more to produce the same amount of electricity? Why worry about efficiency?

Efficiency matters if space is limited. If meeting your electrical needs requires a 3 kW system and the only place with good solar access you have to mount the modules is on a small garage roof, you may need to install the more efficient — and expensive — modules.

While conversion efficiency may not matter to you, efficiency is driving the market, and over the long haul it will result in more efficient and hopefully less expensive modules. Those companies that can produce higher efficiency modules at a lower cost with less raw material in an environmentally friendly manner stand to make billions.

CHAPTER 4

SOLAR SITE ASSESSMENT

Before you invest your hard-earned money in a solar electric system, it is important to determine whether a solar electric system makes sense for your particular situation. Will a PV system meet your needs? How much will it cost? Would you be better off using utility power? Or would some other type of renewable energy system make more sense? And even if it doesn't make perfect sense economically, what about the environmental benefits and personal gratification of doing the right thing? What about alternatives to buying your own PV system, in particular, leases and power purchase agreements?

In this chapter, we begin by focusing on whether a PV system makes sense from an economic perspective. The decision to invest time and money in a PV system requires an analysis of three key factors: (1) how much electricity you need, (2) how much of your electrical demand a PV system can satisfy, and (3)

perhaps most important, how much it will cost to meet your needs with solar electricity.

Assessing Electrical Demand

One of the most common questions we're asked is "How much is it going to cost to install a solar electric system?"

The answer is, "That depends."

The cost of a solar electric system depends on many factors, among them the amount of sunlight you receive, the size of the system, the complexity of the system, the distance the installer must travel, the type of installation, and the difficulty of the installation.

Because solar resources and electrical consumption vary, most residential solar electric systems fall within the 1 to 6 kW range, the most common being 3 to 6 kW systems. For example, Dan's household is extremely energy efficient, consuming 75 to 80% less electricity than a standard home its size, so his system is only 1.1 kW. Ed Begley, Jr., environmental advocate and actor, on the other hand, powers his small home and an electric car in California with a 6 kW system.

In an all-electric home equipped with a wide assortment of electric appliances — especially heavy hitters like central air conditioners, electric space heaters, electric water heaters and electric stoves — average monthly electrical consumption typically falls within the 2,000 to 3,000 kilowatt-hours/month range. Even in very sunny climates, very large PV systems are required to meet their electrical demand. Even a 10 kW system might not be sufficient in such cases.

In homes in which natural gas or propane is used to cook food, heat water, and provide space heat, electrical consumption may be as low as 400 or 500 kWh per month. A much smaller system would be needed in such instances. In a relatively sunny climate, a 3 to 4 kW system might suffice.

Although many factors affect the cost of a PV system, as just noted, the size of the system depends on electrical consumption and solar resources. How you go about calculating

Average Annual Electrical Use

Based on national averages, US homes consume about 1,000 watts continuously, or about 24 kWh per day. That's 730 kWh per month or 8,760 kWh per year, which is about $876 worth of electricity a year (not counting meter-reading fees, taxes and surcharges) at 10 cents per kilowatt-hour.

the consumption depends on whether it is an existing structure or one that's about to be built. Let's begin with existing structures.

Assessing Electrical Demand in Existing Structures

Assessing the electrical consumption of an existing home or business is fairly easy. Most people obtain this information from their monthly electric bills, going back two to three years, if possible. (In some areas, every utility bill includes a summary of year-to-date electrical consumption.) If you don't save your electric bills, a telephone call to the local power company will usually yield the information you need. Most utilities are happy to provide data on electrical energy use to their customers. Some allow customers to access the data online through their websites. All you need is your customer number.

If you purchased a home that's been around for a while, you can obtain energy data from the previous owner. If they've saved their utility bills, they may be willing to share this information with you. Or, they may be willing to contact the local utility company on your behalf to request a summary of their electrical consumption over the past two to three years. Remember, however, a house does not consume electricity, its occupants do, and we all use energy differently. If a previous homeowner and his or her family used energy wastefully, their energy consumption data may be of little value to you if you and your family are energy misers.

To determine total annual electrical consumption from utility bills, don't look at the cost in dollars and cents, look instead for the kilowatt-hours of electricity consumed each month. Calculate the total for each year and then calculate a yearly average. If you're considering installing a grid-tied solar electric system — a PV system that feeds surplus electricity back to the grid — annual electrical demand is all you need. You can size your system based on this information. (We'll show you how, shortly.)

If you are thinking about severing your ties with the utility — that is, taking your house off the grid, a step we don't typically recommend — you will need to calculate average annual consumption and monthly averages, that is, how much electricity is used, on average, during each month of the year. To do this, add up household electrical consumption for each month, and then divide by the number of years' worth of data you have. For example, if your records go back four years, add the electrical consumption for all four Januarys, and then divide by four. Do the same for February and each of the remaining months. Record monthly averages on a table. This will permit you to determine when electrical demand is the greatest. Off-grid systems are sized to meet demands during the times of highest consumption.

After you have calculated annual energy consumption, in the case of a grid-connected PV system, or monthly averages, in the case of an off-grid system, take a few moments to look for trends in energy use. Is energy consumption on the rise or is it staying constant or declining? If you spot an increase in electrical use in recent years, for instance, because

you've installed a big screen TV or added air conditioning, you may want to dump earlier energy data and recalculate the averages based on the most recent bills. Early data will artificially lower the average and the averages won't represent current consumption. You could end up undersizing your PV system.

If, on the other hand, electrical energy consumption has declined, because you've replaced your inefficient furnace and energy-hog refrigerator with newer more frugal models, earlier data will artificially inflate electrical demand. You'll need less electricity than the averages indicate and a smaller system.

Assessing Electrical Demand in New Buildings

Determining monthly electrical consumption is fairly straightforward in existing homes and businesses. Determining electrical demand in a new home — either one that's just been built or one that is about to be built — is much more difficult.

One method used to estimate electrical consumption is to base it on the electrical consumption of your existing home or business. For instance, suppose you are building a brand new home that's the same size as your current home. If the new home will have the same amenities and the same number of occupants, electrical consumption could be similar to your existing home.

If, however, you are building a more energy-efficient home and are installing much more energy-efficient lighting and appliances and incorporating passive solar heating and cooling (all of which we highly recommend), electrical consumption could easily be 50%, perhaps 75% lower than in your current home. If that's the case, you can adjust electrical demand to reflect the efficiency upgrades.

Another way to estimate electrical consumption is to perform a load analysis. A load analysis is an estimate of electric consumption based on the number of electronic devices in a home, their average daily use, and energy consumption. To perform a load analysis, a homeowner begins by listing all the appliances, light and electronic devices in his or her new home or office. Rather than list every light bulb separately, however, you may want to lump them together by room. Dan likes to work with clients one room at a time, using a spreadsheet he prepares for each project. Or, you can use a worksheet like the one shown in Table 4-1. They can be found online at the Solar Living Institute's web page or may be provided by your PV installer.

Once you've prepared a complete list of all the devices that consume electricity, your next assignment is to determine how much electricity each one uses.

The amount of electricity consumed by an appliance, lamp or electronic device can be determined by consulting a chart like the one in Table 4-2. Charts such as this list typical wattages for a wide range of electrical devices, including stoves, microwaves and stereos. Detailed listings are available online. The website of WE Energies, for instance, posts a chart of the approximate electrical consumption of appliances and electronic devices. (You can find this chart by visiting webapps.we-energies.com/

appliancecalc/appl_calc.cfm.) Similar data is available in several books, among them Solar Energy International's *Photovoltaics: Design and Installation Manual* or John Schaeffer's *Solar Living Source Book*.

For more accurate data, check out the nameplate on the appliances or electronic devices you'll be including in your new home or business. You'll find a sticker or metal plate on these devices that lists the unit's rated volts and amps or volts and watts. Voltage may be listed as 120 volts, 120 V, 120 volts AC or 120 VAC. They are all the same. If the nameplate lists volts and amps, but not watts, you will have to calculate watts. To do so, simply multiply the amps x volts (amps x volts = watts).

Multiplying amps times volts yields the wattage of many electronic devices, among them resistive devices, such as electric heaters and electric stoves. (A resistive device is one in which electricity flows through a metal that resists the flow of electrons. This, in turn, produces light and heat.) It also works for universal motors. Universal motors are typically found in smaller devices such as vacuum cleaners, blenders and small electric tools. For devices with induction motors, such as fans, washing machines, clothes dryers, dishwashers, pumps and furnace blowers, multiplying amps by volts significantly overestimates the wattage. For these devices, multiply watts (the product of amps and volts) by 0.6 to obtain a better estimate of true wattage. (For a description of the rationale behind this derating, see the accompanying sidebar, "Real Power, Apparent Power and Power Factor.")

Some experts reduce the power requirements of other devices as well, for example, televisions, stereos and power tools. Why?

Energy auditors reduce the wattage of these devices because the listed wattage is not representative of the typical run wattage — that is, the wattage an appliance will draw when in normal operation. Put another way, the wattage listed on an appliance nameplate, is the power the device draws at maximum

Table 4.1 Electrical Consumption Chart											
Individual Loads	Qty	X Volts	X Amps	= Watts AC	DC	X Use Hrs/day	X Use days/wk	÷ 7 days	= Watts AC	Hours DC
								7		
								7		
								7		
								7		
								7		
								7		
								7		
								7		
								7		
								7		
								7		
								7		
								7		
								7		
								7		
								7		

AC Total Connected Watts: _____

DC Total Connected Watts: _____

AC Average Daily Load: _____

DC Average Daily Load: _____

Table 4.2
Average Electrical Consumption of Common Appliances

General household

Air conditioner (1 ton) 1500
Alarm/Security system 3
Blow dryer...............................1000
Ceiling fan...............................10-50
Central vacuum........................750
Clock radio 5
Clothes washer......................1450
Dryer (gas)................................300
Electric blanket 200
Electric clock..................................4
Furnace fan 500
Garage door opener...............350
Heater (portable)....................1500
Iron (electric)...........................1500
Radio/phone transmit 40-150
Sewing machine........................100
Table fan 10-25
Waterpik 100

Refrigeration

Refrigerator/freezer 540
  22 ft³ (14 hrs/day)
Refrigerator/freezer 475
  16 ft³ (13 hrs/day)
Sun Frost refrigerator..............112
  16 ft³ (7 hrs/day)
Vestfrost refrigerator/ 60
  freezer 10.5 ft³
Standard freezer......................440
  14 ft³ (15 hrs/day)
Sun Frost freezer......................112
  19 ft³ (10 hrs/day)

Kitchen appliances

Blender 350
Can opener (electric) 100
Coffee grinder 100
Coffee pot (electric)...............1200
Dishwasher.............................1500
Exhaust fans (3)........................144
Food dehydrator......................600
Food processor.........................400
Microwave (.5 ft³) 750
Microwave (.8 to 1.5 ft³)........1400
Mixer 120
Popcorn popper 250
Range (large burner).............2100
Range (small burner).............1250
Trash compactor 1500
Waffle iron...............................1200

Lighting

Incandescent (100 watt) 100
Incandescent light 60
  (60 watt)
Compact fluorescent 16
  (60 watt equivalent)
Incandescent (40 watt)..............40
Compact fluorescent 11
  (40 watt equivalent)

Water Pumping

AC Jet pump (¼ hp)..................500
  165 gal per day, 20 ft. well
DC pump for house 60
  pressure system (1-2 hrs/day)

DC submersible pump 50
  (6 hours/day)

Entertainment

CB radio.......................................10
CD player.....................................35
Cellular telephone 24
Computer printer.....................100
Computer (desktop) 80-150
Computer (laptop)................20-50
Electric player piano 30
Radio telephone 10
Satellite system (12 ft dish) 45
Stereo (avg. volume) 15
TV (12-inch black & white).........15
TV (19-inch color)......................60
TV (25-inch color).....................130
VCR 40

Tools

Band saw (14")........................1100
Chain saw (12")......................1100
Circular saw (7¼") 900
Disc sander (9")......................1200
Drill (¼")...................................250
Drill (½")...................................750
Drill (1")..................................1000
Electric mower 1500
Hedge trimmer 450
Weed eater 500

Real Power, Apparent Power and Power Factor

To calculate power consumption (watts) of an electrical resistance device, such as electric stoves and incandescent lamps, simply multiply the voltage by the amps (power [watts] = volts x amps). However, for devices containing inductors or capacitors, such as induction motors, transformers and fluorescent lamp ballasts, the product of volts and amps is actually a bit higher than the actual wattage. In such instances, watts x volts is known as the *apparent power.* The discrepancy is due to the fact that inductors and capacitors store energy as *current,* and voltage waveforms *alternate* from positive to negative. This stored energy causes the current waveform to be out of step with the voltage waveform. When the current waveform is out of step with the voltage waveform, more current is required to deliver the same amount of power. The factor that accounts for this difference between apparent power and real power is the *power factor.*

Power factor is the real power divided by the apparent power. Power factor = watts / (volts x amps). Power factor is always a number between zero and one. When calculating the real power (watts) for a device that has inductors or capacitors, you must multiply voltage times amps times power factor. Power = volts x amps x power factor. Unfortunately, you won't find power factor listed on appliance nameplates, but a reasonable estimate for most household devices containing inductors or capacitors is 0.6.

load. Most devices rarely operate at maximum load, so the nameplate power overstates the actual power used. Reducing the nameplate watts by 25% is a reasonable adjustment for such devices.

Be sure when determining the run wattage of laptop computers that plug into a charger or transformer to use the wattage listed on the charger, not the device. The same applies to cordless drills, cordless phones and electronic keyboards.

A more accurate way to determine the wattage (power) of an electronic device is to measure it directly using a meter like those shown in Figure 4-1. To use a meter such as these, plug the meter into an electrical outlet and then plug the appliance into the outlet on the face of the meter. A digital readout indicates the instantaneous power (watts). When measuring a device that cycles on and off, such as a refrigerator, or one that has a varying load, leave the watt-hour meter connected for a day or two. The meter will record the total energy used during this period in watt-hours. You can then calculate an average over that period.

After you have determined the wattage of each electrical device and light, either by direct measurement or by consulting the nameplate or by a combination of the two, you must estimate the number of hours each one is used on an average day and how many days each device is used during a typical week. From this information, you calculate the weekly energy consumption of all devices in your home or business. You then divide this number by seven to determine the average daily consumption of your home or business in watt-hours. You will use this number to determine the size of your PV system.

Load analysis may seem simple at first glance, but it is fraught with problems. One shortcoming we've encountered is that it is often difficult for clients to estimate how long each device runs on a typical day. Most people have trouble estimating how many hours

Fig. 4-1: *Kill-A-Watt and Watts Up? Meters. (a) The Kill-A-Watt meter and (b) Watts Up? meter shown here can be used to measure wattage of household appliances and electronic devices. Of the two, the Watts Up? is the more sensitive and allows for measurement of tiny phantom loads as well.*

a   b

Plug-in Watt-Hour Meters

The Kill A Watt meter or the Watts Up? meter make it easy to measure power (watts) and energy (watt-hours) used by small devices. Just plug the meter into an outlet and plug the appliance into the meter. These meters read real power (watts), apparent power (volt amps), power factor, volts, amps, and elapsed time (used for calculating average power use over time).

Power meters are useful for estimating load, and also for gaining an appreciation for the energy demands of household appliances. Knowing which appliances and devices are the energy hogs in a home may inspire occupants to reduce energy consumption. Surveying the energy use of household appliances could be a great educational project for children.

Both meters are available in a range of models, some with advanced features. The basic model Kill A Watt meter list price is about $35; the basic model Watts Up? meter list price is about $96. As an alternative to purchasing a meter, some public libraries and utilities have plug-in watt-hour meters available for loan.

or minutes they operate their toasters or blenders each day. Is it three minutes, five minutes, or ten? While some may be able to provide a fairly accurate estimate of how long the kitchen light is on, they haven't the foggiest idea how long the refrigerator runs or how many minutes they run the microwave. And what about the kids? Do they leave the lights on when you're not around? How many hours of television do they watch each day? How many hours a day do they spend on their computers?

Another problem with this process is that run times vary during the year by season. Electrical lights, for example, are used much more in the winter, when the days are shorter, than in the summer. Furnace blowers operate a lot during the winter, but not at all or very infrequently the rest of the year. So what's the average daily run time for the furnace blower?

Another problem with this approach is that many electronic devices draw power when they're off. Such devices are known as a *phantom loads.* They include television sets, VCRs, satellite phone receivers, cell phone and laptop computer chargers, and a host of other common household devices. A few, like satellite receivers, draw nearly as much power when they're off as when they're on (Figure 4-2). Phantom loads typically account for 5 to 10% of the monthly electrical consumption in US homes. If they're not factored into the load analysis, estimates can be off.

Because of these problems, homeowners often grossly underestimate their consumption of electricity. So, if you use this technique, include phantom loads and be generous with your estimates. It's best to overestimate electrical demand, especially if you're sizing an off-grid system.

Fig. 4-2: *Phantom Load. Unbeknownst to most of us, our homes are filled with phantom loads, devices that draw power when not in use, like this television set and satellite receiver. Of this duo, the satellite receiver is by far the worst offender. It uses 15 watts when on and 14 watts when turned off. Phantom loads account for about 5 to 10% of a home or business' annual electrical demand.*

Measuring Phantom Loads

Phantom loads may be quite small, but they add up over time. Plug-in watt-hour meters can be used to measure energy consumed by phantom loads — if you let them run long enough for the kWh reading to register on the display. The Kill A Watt meter displays kWh to two decimal places (10 watt-hours resolution) and the Watts Up? meter displays kWh to one decimal place. A device that draws 5 watts when off, if monitored for 24 hours, would draw 120 watt-hours. This would display as 0.12 kWh on the Kill A Watt meter and would display as 0.1 on the Watts Up? meter.

Once you've calculated daily electrical energy use, it's time to size a solar electric system, right?

Actually, no. Before you size a system, it's a good idea to look for ways to reduce electrical use through energy-efficiency and conservation measures. Why? The lower the energy demand, the smaller a solar electric system you'll need to install. The smaller the system, the less you'll spend to purchase and install your system. How much can you save?

Richard Perez, founder of *Home Power* magazine asserts that every dollar invested in energy efficiency saves $3 to $5 in the cost of a solar electric system. For instance, if you invest $1,000 in measures that trim the electrical energy use in your home or office, for instance, by sealing leaks, adding insulation, and installing more efficient lighting, you could save $3,000 to $5,000 in the cost of a solar electric system.

Bob tells his clients "It's easier and much cheaper to save a watt than to make a watt."

Home Power magazine's Ian Woofenden reminds us that "eliminating energy waste is also a smart move for the environment. While the fuel for renewable energy systems is available daily at no charge and no environmental impact, the collection equipment requires natural resources and the manufacturing processes have an environmental impact. The smaller your system is, the less impact it will have."

Money spent on energy efficiency measures also has a much better return on investment than money spent on a larger generating system, as illustrated in the example shown in Table 4-3, provided by Kurt Nelson, a veteran solar installer and instructor at the Midwest Renewable Energy Association. In this example, the homeowner takes four steps to reduce her energy consumption by 3,552 watt-hours per day. As shown, she replaces her refrigerator and freezer with more efficient models and installs six compact fluorescent light bulbs. She also takes

Table 4.3 The Cost of Energy Measures and Energy Savings				
Efficiency Measure	**New Consumption**	**Old Consumption**	**Energy Savings**	**Cost to implement**
New Refrigerator	1,300 Wh/day	2,200 Wh/day	900 Wh/day	$849
New Chest Freezer	900 Wh/day	1,800 Wh/day	900 Wh/day	$799
6 – 18 Watt CFLs	432 Wh/day (@4 hr/day)	1,800 Wh/day (6 - 75 w @ 4hrs/day)	1,368 Wh/day	$24
Phantom Loads	0	384 Wh/day	384 Wh/day	$13

steps to reduce phantom loads, for instance, installing power strips into which she plugs computers, televisions and stereo equipment so they can be shut off completely when not in use. The total cost was $1,685 (Table 4.4). The annual savings was $92 per year, producing a return on investment of 5.5% per year.

To produce the same amount of energy, the homeowner would need to install a 1 kW solar electric system. The cost of the system is $10,000. It would save the homeowner $92 each year and the return on investment is a paltry 0.9%. Even if the homeowner received 30 to 50% for various rebates and incentives, the system would cost considerably more than efficiency measures.

Because efficiency is both economically and environmentally superior to increasing the capacity of a renewable energy system, most reputable solar installers recommend energy conservation and energy efficiency measures first. If you agree to pursue these measures, they downsize your system and save you thousands of dollars. How do you determine the most cost effective measures to reduce electrical energy consumption?

One way is to hire a professional solar site assessor. For $300 to $400, they will assess the solar potential of your site for a PV system, recommend placement of the system, provide a list of ways you can slash your energy demand, and size your system. They typically rank energy conservation and energy efficiency measures in order of priority. Those that reap the greatest benefits at the least cost are highest on the list. (To locate a solar PV

site assessor in your area, visit the Midwest Renewable Energy Association's website.)

Another option is to hire a home energy auditor. A qualified home energy auditor will perform a more thorough energy analysis of your home and make recommendations on ways you can reduce your demand. They will also prioritize energy-saving measures. Or, you can perform your own home energy audit. For guidance, you may want to check out one of Dan's newest books, *Green Home Improvement*.

Conservation and Efficiency First!

Virtually all renewable energy experts agree: before you invest in a renewable energy system,

Table 4.4 Energy and Economic Savings from Efficiency Measures	
Daily Energy Savings	3,552 watt-hours per day
Annual Energy Savings	1,296 kilowatt-hours per day
Annual Cost Savings	$92
Cost of the Improvements	$1,685
Return on Investment	5.5%

Table 4.5 Comparison of Costs and Savings from Efficiency vs. PVs		
	Energy Efficiency	PV System
Cost	$1,685	$10,000
Annual Savings	$92	$92
Return on Investment	5.5%	0.9%

such as solar electricity, the first step should be to make your home — and your family — as energy efficient as possible. Even if you're an energy miser, you may be able to make significant cost-saving cuts in energy use.

Waste can be slashed many ways. Interestingly, though, the ideas that first come to mind for most homeowners tend to be the most costly: energy-efficient washing machines, dishwashers, furnaces, air conditioners, and our perennial favorite, new windows (Figure 4-3). While vital to creating a more energy-efficient way of life or business, they're the highest

Before a homeowner invests a single penny in a renewable energy system, such as solar electricity or wind, the first step should be to make their home as energy efficient as possible.

fruit on the energy-efficiency tree — and thus the hardest and most expensive.

Before you spend a ton of money on new appliances or better windows, we strongly recommend that you start with the lowest-hanging fruit. These are the simplest and cheapest improvements and yield the greatest energy savings at the lowest cost.

Huge savings can be achieved by changes in behavior. You've heard the list a million times: turning lights, stereos, computers and TVs off when not in use. Turning the thermostat down a few degrees in the winter and wearing sweaters and warm socks. Turning the thermostat up in the summer and running ceiling fans. Opening windows to cool a home naturally, especially at night, and then closing the windows in the morning to keep heat out during the day. Drawing the shades or blinds on the east, south, and then west side of your house as the summer sun moves across the sky can help reduce cooling costs. All these changes cost nothing, except a little of your time, but can reap enormous savings — not just in your monthly energy bill, but also in the cost of a PV system.

There's other low-hanging fruit, too, for example, boosting insulation levels, and weather-stripping and caulking to seal leaks in the building envelope — tightening up and bundling up our homes and workplaces. These measures reduce heat loss in the winter and heat gain in the summer. These steps not only reduce fuel bills, they dramatically increase comfort levels.

Once you've tackled these simple, relatively inexpensive measures, it's time to consider

Fig. 4-3: *Efficiency First! Spending a little extra for an efficient front-loading washing machine like this one will not only save on water and energy over the lifetime of the appliance, but will also reduce the size of your solar system, saving upfront as well.*

more costly, big-ticket items. They, too, can make a huge dent in your monthly energy use. One big-ticket item is your refrigerator. In many homes, refrigerators are responsible for a staggering 25% of the total electrical consumption. If your refrigerator is old and in need of replacement, unplug the energy hog and replace it with a new, more energy-efficient model.

Thanks to dramatic improvements in design, refrigerators on the market today use significantly less energy than refrigerators manufactured 15 years ago. Whatever you do, don't lug the old fridge out to the garage or take it down to the basement and use it to store an occasional case of soda or beer. It will rob you blind!

Waste can also be reduced by replacing energy-inefficient electronics with newer, considerably more efficient Energy Star televisions, computers and stereo equipment (Figure 4-4). You can also trim some of the fat from your energy diet by installing more energy-efficient lighting, such as compact fluorescent lights. (For more on Energy Star appliances see the accompanying sidebar, "Energy Star — and Beyond.")

Although efficiency has been the mantra of energy advocates for many years, don't discount its importance just because the advice has grown a bit threadbare. As it turns out, very few people have heeded the persistent calls for energy efficiency. And many who have made changes have not fully tapped the potential savings.

Readers interested in learning more about making their homes energy efficient may

Fig. 4-4: *Energy Star. When shopping for electronic devices and appliances look for the Energy Star label.*

want to read the chapters on energy conservation in one of Dan's books, for example, *Green Home Improvement*, or *The Homeowner's Guide to Renewable Energy*. Another valuable book is the *Consumer Guide to Home Energy Savings*; and *Home Energy* magazine is also very useful. All these contain numerous tips on making homes much more efficient, more comfortable, and more economical to operate.

Sizing A Solar Electric System

Once you have estimated your electrical demand and implemented a strategy to use energy more efficiently, it is time to determine the size of the system you'll need to meet your needs. This is a step usually performed by professional solar electric installers. The size of the system varies, of course, depending on the type of system you plan to install. Although we'll elaborate on system options in the next chapter, for now, it's important to note that PV systems fall into three broad categories: (1) grid-connected, (2) grid-connected with battery backup, and (3) off-grid.

A grid-connected or utility-connected solar electric system can be sized to meet some or all of the electrical needs of a home or business. Excess electricity, if any, is fed back onto the grid, running the electric meter backward. These systems are by far the most popular. As you will see in Chapter 5, however, when the utility grid goes down, the

Energy Star — and Beyond

Although all new appliances and electronics must comply with federal energy-efficiency regulations, those that exceed these standards display an Energy Star label. This label indicates that the appliance or electronic you're looking at is one of the most efficient in its product category. So, when it is time to replace a television or stereo — even a cordless phone or laptop computer — always look for models with the Energy Star label. As Table 4-6 shows, Energy Star refrigerators use 20% less energy than is federally mandated. Dishwashers use 41% less and clothes washers use 37% less energy than federal law requires.

Although buying an Energy Star rated appliance or electronic device is a good idea, you can do better, sometimes much better. How?

Go to the Energy Star website and click on the appliance or electronic device listing. Here you'll find a list of all Energy Star qualified models. The list indicates the percentage by which each device exceeds the minimum efficiency for that type of appliance.

As Table 4-6 shows, the best appliances on the list are considerably more energy-efficient than other Energy Star qualified products.

If you shop from the Energy Star list, you will do well, but if you search the list for the most efficient models, you can do much better.

For an up-to-date list of energy-efficient appliances, US readers can log onto the EPA's and DOE's Energy Star website at www.energystar.gov. Click on appliances. Or, *Consumer Reports* has an excellent website that lists energy-efficient appliances as well. Their site also rates appliances on reliability, another key factor. Canadian readers can log on to oee.nrcan.gc.ca/energystar/index.cfm for a list of international Energy Star appliances.

Table 4.6 Energy Star Performance		
Appliance Type	**Energy Star Criteria** (% better than federal mandated efficiency)	**Best on the List**
Refrigerators	20%	53%
Dishwashers	41%	147%
Clothes Washers	37%	121%

solar system also shuts down, even on sunny days. (This feature prevents electricity from being fed onto a malfunctioning grid, which might be hazardous to utility company employees.)

A grid-connected system with battery backup is similar to a grid-connected system, but has a battery bank to store electricity. These systems will continue to supply critical loads if the grid goes down.

As their name implies, off-grid PV systems are not connected to the utility grid. They're completely autonomous electric-generating systems. Surpluses are stored in batteries for use at night or on cloudy days.

Sizing a Grid-Connected System

Sizing a grid-connected system is the easiest of all. To meet 100% of your needs, simply divide your average daily electrical demand (in kilowatt-hours) by the average peak sun hours per day for your area. As noted in Chapter 2, peak sun hours can be determined from solar maps and from various websites and tables (also, see sidebar in this chapter, "Determining Peak Sun Hours").

As an example, let's suppose that you and your family consume 6,000 kWh of electricity per year (or 500 kWh per month). This is about 17 kWh per day. Let's suppose you live in Lexington, Kentucky where the average peak sun hours per day is 4.5. Dividing 17 by 4.5 gives you the array size (capacity), 3.8 kilowatts. But don't run out and order a system based on that number.

This calculation yields system size if the system were 100% efficient and unshaded throughout the year. Unfortunately, no PV system is 100% efficient. As a result, most solar installers derate grid-connected PV systems by 25 to 30%. This loss is due to voltage drop as electricity flows through wires; resistance at fuses, breakers and connections; dust on the array; and inefficiencies of various components such as the inverter. So, if your PV array is not shaded by trees or nearby buildings, you'd need a 25 to 30% larger system to provide 17 kWh of electricity per day. We'll use 30%. To calculate the array size you would need, divide 3.8 kW by 0.70 to account for inefficiencies. The result is 5.4 kW. Let's round it up to 5.5 kW to be on the safe side. To meet your needs, then, you would need a 5.5 kW system *provided the solar array is not shaded.*

Shading dramatically lowers the output of a PV system. Even a small amount of shade on a module can reduce its output — quite dramatically. To determine the amount of shading on a PV system, professional solar site assessors and solar installers often use a sun path analysis tool like the Solar Pathfinder, which is shown in Figure 4-5. The Solar Pathfinder and similar devices determine the percentage of solar radiation blocked by permanent local features in the landscape such as trees, hills and buildings. These devices are used by solar installers to locate the sunniest (most shade-free) site on a property and, then, to determine how much shading, if any, will occur at that location throughout the year.

The Solar Pathfinder consists of a clear plastic dome over a grid that indicates the location of the Sun throughout the year at

Fig. 4-5:
Solar Pathfinder.
This device allows
solar site assessors
and installers to
assess shading at
potential sites for
PV systems, helping
you select the best
possible site to
install an array.

SHAWN SCHREINER

Fig. 4-6:
The dome of the
Solar Pathfinder
shows all sources of
shade throughout
the year, no matter
when you use the
device. This allows
solar site assessors
to determine how
much shading
occurs and when,
so an array's output
can be accurately
estimated.

SHAWN SCHREINER

any given location (Figure 4-6). The Solar Pathfinder is set on a tripod at the proposed location of a PV array, pointed due south (true south), and leveled. Once set up, reflections of trees and other objects that could shade a PV array show up on the glass dome.

To estimate shading, the operator can trace the reflections showing up on the dome on the Sun path diagram beneath the dome, and then manually calculate the percent of sunlight that occurs each month. Or, the operator can take a photo of the reflections on the dome with a digital camera. The digital image is then downloaded into accompanying software, Solar Pathfinder Assistant. It calculates the percent of sunlight for each month of the year. The amount of sunlight on the proposed array is then used to adjust the size of the array. (The percentage of sunlight can also be used to predict the output of an array whose size was determined by budget — for example, for an individual who only has enough money to install a 2 kW array.)

To calculate the actual output of an array, the Solar Pathfinder Assistant software merges the data with another software program, known as PV Watts. This software, which was developed by the National Renewable Energy Laboratory (NREL) in Golden, Colorado, uses historic weather data to calculate an array's output based on shading from clouds. It is available online for free. (On its own, however, it does not compensate for shading by obstacles. That's why the Solar Pathfinder Assistant software or other similar software must be used.)

To see how this works, consider an example. Table 4-7 shows the output of a 5 kW

Solar Resource Evaluation and Shading Tools

In the text, we describe the use of the Solar Pathfinder, the first solar resource evaluation and shading tool. Since its debut, two additional solar resource evaluation tools have been introduced into the market, ASSET(Acme Solar Site Evaluation Tool), manufactured by Wiley Electronics (www.we-llc.com); and the Solmetric SunEye manufactured by the Solmetric Corporation (www.solmetric.com).

ASSET and SunEye are more sophisticated tools, more automated, and more costly. While Solar Pathfinder costs about $250, ASSET costs about $600 and SunEye costs about $1,400.

ASSET consists of a tripod-mounted digital camera (Figure 4-7). To set it up, the camera is attached to the aluminum tripod, leveled, and then oriented to the south. A bubble level and compass are mounted on the assembly. Seven to nine photos are then taken, scanning a potential solar site from the east to the west. The images are downloaded into a computer (either a PC or a Mac), where they are "stitched together" electronically to produce a single panoramic view. The accompanying software then superimposes a sun path grid, like those used in the Solar Pathfinder. The grid shows the sun path during the entire year at a specific latitude. The software then automatically calculates total solar shading. It uses this data to calculate the available insolation.

The SunEye, shown in Figure 4-8, is a handheld device. It consists of a Hewlett-Packard PDA, software, a built-in digital camera, a bubble level, and a compass. The camera takes a panoramic picture of the site, instantly superimposes the sun path, and calculates shading (actually available sunlight, called *solar access*). It generates a bar graph that shows the monthly solar access. This information can then be downloaded onto a computer to generate reports or to be archived for later use. The software that comes with SunEye, like ASSET's software, permits the user to edit out trees — and other objects — that shade an array to determine the gains that would be created by clearing a site of trees or other objects that might block the sun.

WILEY ELECTRONICS

Fig. 4-7: *ASSET*

SOLMETRIC CORPORATION

Fig. 4-8: *Solmetric SunEye*

array in a location with average peak sun hours of 4.5 (near Lexington, Kentucky). This data is from NREL's PV Watts program, available online. Column 1 lists the months of the year. Number 1 is January. Column 2 lists the solar radiation, or average peak hours per day by month. It's shown in kWh/m$_2$/day. The yearly average is shown at the bottom of the column. Column 3 shows the amount of electricity a 5 kW array will produce each month. The total output is listed at the bottom of the column. Column 4 shows the value of the electricity at the current rate of 6.1 cents per kWh. As indicated in the bottom of the column, a 5 kW array will produce about $365 worth of electricity annually.

The output is then adjusted by Solar Pathfinder Assistant. Table 4-8 indicates the monthly output based on shading. Column 4 is the percent sun determined by the solar pathfinder. Column 5 is the AC output of an array based on shading.

If the family required 6,000 kWh per year, the array size would need to be increased. To adjust the array size, divide 6,000 (the annual electrical consumption) by 5,632 (6,000/5,632 = 1.06) and multiply the result (1.06) by 5. In the case shown in Table 4-8, the homeowner would need a 5.3 kW PV array to meet his or her needs (1.05 x 7.5 = 7.9). This system would cost about $53,000 without the federal tax credit (assuming a grid-connected system costs $10 per watt installed). Taking into account the 30% federal tax credit for homes and businesses starting in 2009, the cost would be $37,100. Cutting electrical demand in demand by half, which would cost about $10,000 in efficiency upgrades, would reduce the system cost to $18,550.

Solar installers can also size a PV system based on budget — how much money a person has to spend — because not everyone can afford a PV system to meet 100% of their needs. For example, suppose a business had $20,000 to invest in a PV system. With the 30% federal tax credit, this would purchase a 2.9 kW grid-connected PV system (at this writing). A 2.9 kW system in an area with an average of 4.6 peak hours of sun per day and no shading would produce about 3,300 kWh of electricity per year. If the company used 20,000 kWh per year, the system would meet about 16.5% of their electrical energy demand.

Month	Solar Radiation (kWh/m$_2$/day)	AC Energy (kWh)	Energy Value ($)
1	3.23	397	24.22
2	3.85	420	25.62
3	4.38	516	31.48
4	5.28	578	35.26
5	5.48	603	36.78
6	5.84	599	36.54
7	5.51	586	35.75
8	5.40	575	35.08
9	4.94	521	31.78
10	4.61	513	31.29
11	3.36	376	22.94
12	2.58	307	18.73
Year	4.54	5991	365.45

Table 4.7
Output of a 5 kW Array in Lexington, KY from PV Watts

DAN CHIRAS

For optimum performance, PV systems generally require a site where an array is unshaded at least three hours on either side of solar noon — that is, from 9 AM to 3 PM solar time. This portion of the daily solar window accounts for 75 to 80% of the sun's daily radiation, depending on the season.

Sizing an Off-Grid System

As a rule, off-grid systems are sized according to the month of the year with the least sunshine. In temperate climates — for example, in states like Wisconsin, Kansas, New York and Oregon — the array is typically sized on insolation in November, December or January. (These are the shortest and typically the cloudiest months of the year in these climates.) First, determine the electrical demand during the month with the lowest insolation. Then divide the average daily consumption by the average peak sun hours during that month to determine the size of the array, then adjust for efficiency and shading, as described previously.

The size of the battery bank must also be carefully calculated. Although we'll discuss battery bank sizing in more detail in Chapter 7, it's important to note that battery banks are sized to provide sufficient electricity to meet a family's or business's needs during cloudy periods. Most battery banks are based on a

Table 4.8 Output of a 5 kW Array based on Shading				
Month	Solar Radiation (kWh/m2/day)	AC Energy (kWh) w/o Shading	% Sun	AC Energy w/Shading
1	3.23	397	82	326
2	3.85	420	90	378
3	4.38	516	90	464
4	5.28	578	95	549
5	5.48	603	100	603
6	5.84	599	100	599
7	5.51	586	100	586
8	5.40	575	100	575
9	4.94	521	100	521
10	4.61	513	95	487
11	3.36	376	90	338
12	2.58	307	90	276
Year	4.54	5991	94	5632

DAN CHIRAS

three to five day reserve so they'll provide enough electricity for three to five days of cloudy weather. Battery banks are also sized to prevent deep discharging during this period — that is, so the batteries do not discharge more than 50% of their capacity. Deep discharging can seriously damage batteries.

Sizing a Grid-Connected System with Battery Backup

PV arrays in grid-connected systems with battery backup are sized much like grid-connected systems, discussed earlier. As a rule, the size of the array is based on expectations, that is, the percentage of electrical demand a system might be required to meet, taking into account peak sun hours, efficiency and shading.

Determining Peak Sun Hours

Solar installers will size your system, but if you're a do-it-yourselfer or an aspiring solar installer, you'll need to size systems on your own. As noted in the text, sizing a system is pretty easy. You simply divide the daily electrical use in kWh by the peak sun hours, and then adjust for efficiency and shading. Where do you find data on peak sun hours?

One source of this data is DOE's State Energy Alternatives website at eere.energy.gov/states/alternatives/resources_mt.cfm. Click on your state to locate the solar map. Use the peak sun hours for flat-plate collectors, which is listed as watt-hours per square meter per day. (Flat-plate collectors are solar collectors with a flat collector surface like PV modules and many types of solar hot water collectors.) Divide by 1,000 to convert this to kilowatt-hours.

Another excellent source is NASA's online database, Surface Meteorology and Solar Energy at eosweb.larc.nasa.gov/sse/. This website provides data on solar energy anywhere in the world. To access the data, however, you'll need to set up a password-accessible account. Don't fret. It only takes a few seconds.

Although the site is very user friendly, here are instructions to help you navigate your way to the data you'll need:

1. Log on to eosweb.larc.nasa.gov/sse/.

2. After logging on to the site, click on Meteorology and Solar Energy.

3. Then either click on "Enter the Latitude and Longitude" or the map of the world to locate your site (the latter is slower).

4. A new screen will appear. It will ask you to enter your e-mail address and password. If you haven't set up an account, do it now.

5. After setting up an account, enter the latitude and longitude of the site.

6. The latitude and longitude of a site can be determined by calling a local surveyor, ☞

Battery banks are sized according to the electrical requirements of *critical loads* — that is, the electrical load that would be needed during a power outage. Critical loads are usually restricted to pumps or fans of heating or cooling systems, well pumps, sump pumps, refrigerators and a few lights. Because these loads require less energy than the entire home, battery are typically much smaller in grid-connected systems with battery backup than off-grid systems.

Does a Solar Electric System Make Economic Sense?

At least three options are available to analyze the economic cost and benefits of a solar electric system: (1) a comparison of the cost of electricity from the solar electric system with conventional power or some other renewable energy technology, (2) an estimate of return on investment, and (3) a more sophisticated economic analysis tool known as discounting.

checking the Gazetteer for your state, referring to a topographic map, or checking out a satellite image of a property online. Aerials that include latitude and longitude are available online at www.terraserver-usa.com.

7. Latitudes and longitudes may be entered as decimal degrees (example 33.5) or as degrees and minutes, separated by a space (example 33 30). Latitudes are either north or south, longitudes are either east or west. Minus signs are assigned to south latitudes and west longitudes to avoid confusion.

8. Once you have entered the latitude and longitude, click on submit.

9. You'll now be presented with a long list of data from which to choose. (You can confirm that you've entered the right latitude and longitude by clicking on the words, "Show a location map." When you do, it will present a map showing the location you've selected. If it's not correct, check the latitude and longitude to be sure you entered it correctly.

10. Scroll down to the first box entitled "Parameters for Tilted Solar Panels." In this box, you can click on one of several options. Click on "Radiation on Equator-Pointed Tilted Surfaces."

11. Scroll to the bottom of the page and click on "Submit."

12. You'll be presented with the data you requested, in this case, Monthly Averaged Radiation Incident On An Equator-Pointed Tilted Surface in kWh/m2/day which is, of course, peak sun hours.

13. This table lists the monthly peak sun hours at different array tilt angles. It also shows the optimum tilt angle for each month and the peak sun hours at the monthly optimum tilt angles.

14. Data tables can be cut and pasted into documents for reports or for your own use later on. ■

Cost of Electricity Comparison

One of the simplest ways of analyzing the economic performance of a solar system is to compare the cost of electricity produced by a PV system to the cost of electricity from a conventional source such as the local utility. This is a five-step process, two of which we've already discussed.

The first step is to determine the average monthly electrical consumption of your home or business, preferably after incorporating conservation and efficiency measures. Second, calculate the size of the system you'll need to install to meet your needs. Third, calculate the cost of the system. (A solar provider can help you with this.) Fourth, after determining the cost of the system, calculate the output of the system over a 20- to 30-year period, the expected life of the system. Fifth, estimate the cost per kilowatt-hour by dividing the cost of the PV system by its total output in kWh.

Suppose you live in Colorado and are interested in installing a grid-connected solar electric system that will meet 100% of your electric needs. Your super-efficient home requires, on average, 500 kWh of electricity per month. That's 16.4 kWh per day. Peak sun hours is 6. To size the system, divide the electrical demand (16.4 kWh per day) by the peak sun hours. The result is 2.7 kW. Adjusting for 70% efficiency, the system should be 3.9 kWh. For the sake of simplicity, let's assume that the system is not shaded at all during the year.

Your local solar installer says she can install the system for $7 a watt or $27,300. You'll receive a rebate from the utility of $3.50 per watt of installed capacity or $13,650. The system cost is now $13,650. You'll also receive a 30% tax credit from the federal government on the cost of the system. The federal tax credit is based on the initial cost of the system ($27,300) minus the utility rebate ($13,650), or $13,650 in this example. Thirty percent of this amount equals $4,095. Total system cost after subtracting this incentive is $9,555.

According to your calculations or the calculations provided by the solar installer, this system will produce about 16.4 kWh of electricity per day or 6,000 kWh per year. If the system lasts for 30 years, it will produce 180,000 kWh.

To calculate the cost per kilowatt-hour, divide the system cost ($9,555) by the output (180,000 kWh). In this case, your electricity will cost 5.3 cents per kWh. Considering that the going rate in Colorado is currently about 9.5 cents per kWh, including all fees and taxes, the PV system represents a pretty good investment.

Economists reading this analysis will object to its simplicity. They'll note that this simple economic comparison does not include important factors such as the cost of borrowing money to purchase a PV system. Interest payments will add to the cost of electricity produced by the system. For those who self-finance by taking money out of savings, this economic analysis fails to take into account opportunity costs — lost income from interest-bearing accounts raided to pay for the system.

To be fair, this simple economic tool also fails to take into account the rising cost of electricity. Nationwide, electric rates have increased on average about 4.4% per year over the past 25 years. In recent years, the rate of increase has been double that amount in some areas.

Although a comparison of the cost of solar electricity to the cost of utility power ignores key economic factors, the rising cost of electricity from conventional sources will in all likelihood offset the opportunity cost or the cost of financing a system.

When calculating the cost of electricity from a solar electric system, be sure to remember to subtract financial incentives from state and local government or local utilities — as in the previous example. Financial incentives can be quite substantial. In Wisconsin, for example, more than 30 utilities participate in a statewide program called Focus on Energy through which they provide customers who install PV systems a rebate of up to 25% of their system cost with a maximum reward of $35,000. Other utilities and even several states, like New York and California, offer generous incentives as well. The best PV incentives are found in Colorado, New Jersey, Massachusetts, California and Oregon.

The federal government also offers a generous financial incentive to those who install PV systems. Their incentive is a 30% tax credit to homeowners and businesses. However, the feds also allow businesses to depreciate a solar electric system on an accelerated schedule, which means they can deduct the costs faster than other business equipment, recouping their investment more quickly. This further reduces the cost of a PV system. The US Department of Agriculture offers a 25% grant to cover the cost of PV systems on farms and rural businesses. Their minimum grant is $2,500 (for a $10,000 system) and the maximum is $500,000. To learn more about state and federal incentives in your area, log on to the Database of State Incentives for Renewables and Efficiency at www.dsireusa.org. Click on the map of your state. To learn more about USDA grant program, log on to rurdev.usda.gov/.

As the previous example showed, financial incentives can reduce the cost of a PV system; consequently, most PV system installations are driven by incentives.

When comparing the cost of an off-grid PV system on a new home to the cost of electricity from the grid, don't forget to include the line extension fees — the cost of connecting the home to the electrical grid. If your new home or business is more than a few tenths of a mile from existing electric lines, you could be charged handsomely to run lines to your home. Remember that the cost of connecting to the electric grid only covers the cost of the installation of poles and electric lines. It does not buy you a single kilowatt-hour of electricity.

Calculating Return on Investment

Another relatively simple method used to determine the cost effectiveness of a PV system is *simple return on investment* (ROI). Simple return on investment is, as its name implies, the savings generated by installing a PV system expressed as a percentage of the investment.

Simple ROI is calculated by dividing the annual dollar value of the energy generated by a PV or wind system by the cost of the system. A solar electric system that produces 6,000 kWh of electricity per year at 9.5 cents per kilowatt-hour generates $570 worth of electricity each year. If the system costs $9,555, after rebates, the simple return on investment is $570 divided by $9,555 x 100 which equals 6%. If the utility charges 15 cents per kWh, the 6,000 kWh of electricity would be worth $900 and the simple ROI would be 9.4%. Given the state of the economy, both of these represent decent rates of return. (If only our retirement funds performed half as well these days!) Even in good economic times, these are respectable ROIs.

As in the cost comparison method, simple ROI fails to take into account interest payments on loans required to purchase the system or opportunity costs, for example, lost interest if the system is paid for in cash. It also fails to take into account system maintenance, insurance or property taxes, if any. All of these factors could decrease the return on investment.

But, as in the analysis based on comparing the cost of electricity, this method fails to take into account a number of factors on the other side of the ledger; for example, rising electricity costs would increase the value of solar-generated electricity. Simple ROI also fails to take into account the fact that money saved on the utility bill is tax-free income to you. There are also possible income tax benefits for businesses, such as accelerated depreciation.

Despite these shortcomings, simple return on investment is a convenient tool for evaluating the economic performance of a renewable energy system. It's infinitely better than the black sheep of the economic tools, *payback* (also known as "simple payback").

Why?

Payback is a term that gained popularity in the 1970s. It was used to determine whether energy conservation measures and renewable energy systems made economic sense. What exactly is payback? Payback is the number of years it takes a renewable energy system or energy efficiency measure to pay back its cost through the savings it generates.

Payback is calculated by dividing the cost of a system by the anticipated annual savings. If the $9,555 PV system we've been looking at produces 6,000 kilowatt-hours per year and grid power costs you 9.5 cents per kWh, the annual savings of $570 yields a payback of 16.8 years ($9,555 divided by $570 = 18.85 years). In other words, this system will take almost 17 years to pay for itself. From that point on, the system produces electricity free of charge.

While the payback of nearly 17 years on this system seems long, don't forget that the return on investment on this system, calculated earlier, was 6%, which we concluded earlier was a very respectable rate of return on your investment — or any investment these days.

While simple payback is fairly easy to understand, this example shows that it has some very serious drawbacks. The most important is that payback is a foreign concept to most of us and, as a result, can be a bit misleading.

Besides being misleading, simple payback is a concept we rarely apply in our lives. Do avid anglers ever calculate the payback on their new bass boats? ($25,000 plus the cost of oil, gas and transportation to and from favorite fishing spots divided by the total number of pounds of edible bass meat at $5 per pound over the lifetime of the boat.) Do couples ever calculate the payback on their new SUV or the new chandelier they installed in the dining room? Did you calculate the payback on your new carpeting?

Interestingly, simple payback and simple return on investment are closely related metrics. In fact, ROI is the reciprocal of payback. That is, ROI = 1/payback. Thus, a PV system with a 10-year payback represents a 10% return on investment (ROI = 1/10). A PV system with a 20-year payback represents a 5% ROI.

Although payback and ROI are related, return on investment is a much more familiar concept. We receive interest on savings accounts and are paid a percentage on mutual funds and bonds — both of which are a return on our investment. Most of us were introduced to return on investment very early in life — when we opened our first interest-bearing account. Renewable energy systems also yield a return on our investment, so it is logical to use ROI to assess its economic performance.

Discounting and Net Present Value: Comparing Discounted Costs

For those who want a more sophisticated tool to determine whether an investment in solar energy makes sense, economists offer up *discounting* and *net present value*. In this method, they compare the cost of a PV system to the cost of the electricity it will displace. Unlike the previous economic tools, discounting and net present value take into account numerous economic parameters, including initial costs, financial incentives, maintenance costs, the rising cost of grid power, and another key element, the time value of money. The time value of money takes into account the fact that a dollar today is worth more than a dollar tomorrow and even more than a dollar a few years from now. Economists refer to this as the *discount factor*.

The discount factor represents something economists refer to as *opportunity cost* and includes inflation. Opportunity cost is the cost of lost economic opportunities by pursuing one investment path over another — for example, investing your money in a solar system instead of investing your money in the stock market.

To make life easier, this economic analysis can be performed by using a spread sheet like the one shown in Table 4-9 provided by renewable energy economist John Richter of the Sustainable Energy Education Institute in Michigan. This spread sheet is available for your use on Dan's website, (evergreeninstitute.org) courtesy of John.

In Table 4-9, the first column is the year. The second column is the discount factor. For simplicity, we recommend choosing the highest interest rate on any debt you have, including your mortgage, as the discount factor. Or, if you have no debt, choose the highest investment interest rate you can get with a risk profile similar to the renewable energy system, which is usually very low. A ten-year government bond is a good basis and currently pays less than 3%. As indicated in column 2, a dollar today will be worth 41 cents in 30 years.

The next column (under the heading "Buy Utility Electricity") shows the cost of electricity from the local power company — that is, how much you will pay each year for electricity purchased from the local utility — taking into account rising fuel costs (4.4% annual increase). As shown here, the PV system produces $570 worth of electricity in year one. Column three shows the cost of electricity per year over a 30-year period had you chosen to buy from the utility. The last entry in this column is the total cost of electricity to you — $34,191. That's how much money you will pay to the utility over the 30-year period if you purchase 6,000 kWh of electricity per year from them,

		Table 4.9			
		Economic Analysis of PV System in Colorado			
Year	**Discount Factor**	**Buy Utility Electricity**		**Proposed PV System**	
	3.0%	**Cost 4.4%**	**Discounted Cost**	**Cost**	**Discounted Cost**
0	1.000	$0	$0	$9,555	$10,710
1	0.971	$570	$553	$0	$0
2	0.943	$595	$561	$0	$0
3	0.915	$621	$569	$0	$0
4	0.888	$649	$576	$0	$0
5	0.863	$677	$584	$0	$0
6	0.837	$707	$592	$0	$0
7	0.813	$738	$600	$0	$0
8	0.789	$771	$608	$0	$0
9	0.766	$804	$617	$0	$0
10	0.744	$840	$625	$0	$0
11	0.722	$877	$633	$0	$0
12	0.701	$915	$642	$0	$0
13	0.681	$956	$651	$0	$0
14	0.661	$998	$660	$0	$0
15	0.642	$1,042	$669	$0	$0

JOHN RICHTER, SUSTAINABLE ENERGY EDUCATION INSTITUTE

based on an inflationary increase of 4.4% per annum.

The next column under the heading "Buy Utility Electricity" is the discounted cost of electricity from the utility. The discounted cost of electricity from the utility is the cost of electricity taking into account the discount rate (the declining value of the dollar due to inflation) applied to the rising cost of electricity. This calculation allows one to calculate "present value" of the money spent on electricity from the utility over a 30-year period. Put another way, that's the value of the money one would spend over a 30-year period in present-day dollars.

As you can see, although you will have shelled out $34,191 to the utility company, the net present value of that money — that is, its value in present-day dollars — is only $20,330.

In the fifth column of the spreadsheet is the cost of the PV system — $27,300 minus incentives, or $9,555. You will note that we added a $3,200 charge after 20 years — that's to replace the inverter. Over a period of 30

		Table 4.9 cont. Economic Analysis of PV System in Colorado				
Year	**Discount Factor**	**Buy Utility Electricity**		**Proposed PV System**		
Rate	**3.0%**	**Cost 4.4%**	**Discounted Cost**	**Cost**	**Discounted Cost**	
16	0.623	$1,087	$678	$0	$0	
17	0.605	$1,135	$687	$0	$0	
18	0.587	$1,185	$696	$0	$0	
19	0.570	$1,237	$706	$0	$0	
20	0.554	$1,292	$715	$3,200	$1,772	
21	0.538	$1,349	$725	$0	$0	
22	0.522	$1,408	$735	$0	$0	
23	0.507	$1,470	$745	$0	$0	
24	0.492	$1,535	$755	$0	$0	
25	0.478	$1,602	$765	$0	$0	
26	0.464	$1,673	$776	$0	$0	
27	0.450	$1,746	$786	$0	$0	
28	0.437	$1,823	$797	$0	$0	
29	0.424	$1,903	$808	$0	$0	
30	0.412	$1,987	$819	$0	$0	
Total		**$34,191**	**$20,330**	**$12,755**	**$9,555**	

JOHN RICHTER, SUSTAINABLE ENERGY EDUCATION INSTITUTE

years, you will have invested $12,755 in your system (in present dollars).

The last column of the spreadsheet is the discounted cost of the PV system. This is the present value of your expenditure, taking into account the discount factor of 3%.

The final step is to compare the discounted cost of the system ($11,327) to the discounted cost of electricity from the utility ($20,330). In this example, the present value of the PV system is $9,003 less than the present value of the cost of utility electricity.

In this analysis, if a PV system is cheaper than buying electricity, it makes economic sense. If it costs more, it doesn't. The greater the difference in the cost of the two systems, the more compelling the decision. Even if the differential is small, however, the investment may be worth it.

Comparisons based on net present value, return on investment, or the costs of electricity are vital to making a rational decision about a PV system. As Richter pointed out at a workshop at the 2006 Michigan Energy

Selling Energy to the Utility — Fact or Fiction?

Many people are enamored of the idea of selling surplus electricity to the utility. Some even imagine that they'll get rich. Is this realistic?

In the previous example of a system that provides 100% of the owner's 500 kWh per month load, the system cost $9,555 after rebates. At the current price of 10 cents per kWh, this homeowner is saving $570 per year. How much is he selling back to the utility?

Nothing.

This system was sized to meet 100% of his demand, so there is nothing left over to sell to the utility. What if the size — and cost — of the system were doubled? Then the owner would produce 500 kWh per month excess energy to sell to the utility. What will the utility pay for this energy?

That depends.

Let's say the customer's utility pays the retail rate. The 500 kW of electricity would be worth about $50 per month, or $600 per year (if the retail cost of electricity is 10 cents per kWh). Not a bad investment.

Unfortunately, only a few states, like Kentucky, Colorado and Minnesota, have such generous policies. Federal law requires utilities to accept surplus energy from its customers (that's surplus electricity at the end of an annual net metering billing cycle, for instance) and to pay for it, but only at their avoided cost. This is the utility's wholesale cost for producing what it sells, which is far less than the retail price it charges its customers. Most states have adopted this pricing in their net metering policies. The avoided cost is usually 2.5 to 4 cents per kWh. At this rate, our homeowner would get a check from the utility for $12.50 to $20 per month.

There could also be additional costs to factor in. When connecting a PV system to the grid, most utilities require the homeowner to install a second electric meter to measure the energy so they can track the flow of electricity to and from the grid. ☛

Fair sponsored by the Great Lakes Renewable Energy Association, "Even though a system may not make perfect sense from an economic standpoint, it is your money. You can spend it how you see fit."

You may want to purchase a PV system for peace of mind — knowing it will free you from utility power and rising fuel costs. Being independent of the local power company is often a compelling motivation. The idea of selling power to the utility may motivate others (even if this is often an illusion — see sidebar,

"Selling Energy to the Utility — Fact or Fiction?"). Or, you may find the personal satisfaction in generating power from a clean, renewable resource sufficient enough reason to invest in a PV system. Or, for some of us, it may be enough just to have a fancy new toy to play with.

Alternative Financing for PV Systems

Some individuals and businesses have the financial wherewithal to pay for their PV systems outright. State, local and federal incentives, of

The meter may cost a few hundred dollars and the utility may also charge $10 or more per month to read the meter.

The homeowner's income is now $2.50 to $10 per month. Not such a good investment.

In some states, like Arkansas, the situation is even worse. The utility simply keeps the surplus electricity at the end of the year. The homeowner receives nothing for his or her surplus.

Net metering laws vary widely across the United States, so be sure to check your state's policy before installing a grid-connected PV system. Some states, like Colorado and Missouri, have extremely favorable policies. In states with a *renewable portfolio standard,* for example, utilities are required to obtain a certain percentage of their energy from renewable sources. If the utility has been mandated to obtain, say, 10% of its energy from renewables, and currently is obtaining only 2%, they may be highly motivated to buy your solar-produced energy. One utility in Wisconsin, for example, pays 25 cents per kWh for renewable energy from small-scale producers. At that rate, our homeowner's income would be $125 per month, minus a $10 meter fee. For a system costing an additional $10,000 the return on investment would be equal to 15%.

Is this a good investment?

Absolutely.

Is the homeowner getting rich?

Not really, but in today's market, he's not doing too bad. With this favorable rate structure, he is (1) getting his household electricity free, and (2) has a modest stream of income that will last for years to come and will grow in value at least to match inflation. Not a bad deal, if you can get it. Unfortunately, there aren't many deals this sweet out there. ■

course, make systems more affordable. Some people take out loans — home equity loans or small business loans — to finance their systems. Unfortunately, not everyone has access to the money required to purchase a system — even with incentives — or wants to incur that kind of debt. If you are one of these people, there are some alternative financing mechanisms that could still make your dreams of a PV system come true: *power purchase agreements* and leases.

Power purchase agreements (PPAs) have been used for decades to finance large commercial power projects. Thanks to the pioneering efforts of California-based SunRun and other similar companies, power purchase agreements are now being used to finance PV systems for homes and small businesses in California.

In an article in *Home Power* magazine (issue 128), Charles W. Thurston, author and market analyst, explains how SunRun's power purchase agreements work. For a fraction of the typical initial cost, he notes, a solar electric system is installed on a customer's home by one of the company's installers. SunRun owns and operates the system and sells the generated electricity to the homeowner at a low rate — usually a rate that increases much more slowly than utility rates for the duration of the lease, typically around 18 years. At the end of each month, the customer receives two bills, one from the utility and the other from SunRun. If the system were sized to meet 100% of a customer's demand, the utility bill would be quite small.

A 2008 law in California makes it easier for solar companies to sell PPAs, so those living in the state can expect to see many other companies entering the field. For those outside of California, SunRun and other companies are looking to expand their programs elsewhere.

Although power purchase agreements have been used to bring PVs to single family homes, they have also been used to finance rooftop PV systems for entire real estate developments in California, thanks to the work of Open Energy Corporation. This company finances the entire project so there's no upfront cost to the developers or future homeowners, according to Thurston. The benefits to developers are several, according to John McCusker, a representative of Open Energy Corporation. Builders receive the tax credits and enjoy robust sales — their units sell four times faster than comparable, non-solar residences in the area. Homeowners benefit, too. Not only do they incur no upfront costs, but they enjoy lower electric bills and live a more environmentally friendly lifestyle. They also own a residence that will probably sell more quickly when the time comes to put it on the market.

Another option, with lower up-front payments, is a lease. In California, SolarCity has teamed up with investment bank Morgan Stanley to provide leases that could make solar electricity even cheaper than it is with most PPAs. The company installs the PV system at their cost. They and the homeowner typically sign a 15-year lease that guarantees a certain amount of electricity from the PV system, based on the size of the system and conditions at the site (tilt angle, orientation and shading). Most systems are slightly undersized, so they primarily produce electricity to offset the most expensive peak utility rates. This lowers lease payments, and maximizes savings for the customer.

Customers that lease PV systems typically end up paying slightly less for electricity. The lease also guarantees a fixed rate for the term of the agreement, providing a hedge against rising electric rates.

Lease programs are available in California, Arizona, Oregon and Connecticut. Expect to see other companies enter the market in other states.

Lease programs and power purchase agreements are really quite similar. The main difference is that in lease programs there's typically no down payment. However, as Thurston explains, "if you can afford to invest up front in part of the system cost (through a PPA), you'll pay less as time goes on, and your savings can be greater at the end of the contract. In that case, a PPA may be more beneficial."

Despite Thurston's analysis, representatives from both industries argue that the financial costs are not that different over the long haul. "The bottom line is that a solar lease or PPA makes it possible for any home-owner to stop talking about tomorrow and act now," says Thurston. If you'd like to power your home with solar electricity, but can't afford a system or don't want to borrow the money, consider a lease or a power purchase agreement. If you can afford a system of your own, and receive generous rebates, you may want to consider installing one yourself. The economic benefits can be many and can result — under optimum conditions — in very affordable and clean electricity.

Putting It All Together

In this chapter, you've seen that there are several ways to save money on a PV system. Efficiency measures lower the initial size and cost of a system, saving huge sums of money. Tax incentives and rebates also lower the cost. Some states exempt PV systems from sales taxes or property taxes, creating additional economic savings. Avoiding line extension fees by installing an off-grid system in a new home rather than a grid-connected system can also save huge amounts of money, often enough to pay for a good portion, or perhaps even all of your system cost.

Dan even encourages those who are building superefficient passive solar/solar electric homes to view savings they'll accrue from efficiency measures and passive solar design as a kind of internal subsidy or rebate for their PV systems. Dan's own solar electric system cost about $17,000 and has generated about $4,000 worth of electricity in the first 13 years. The return on investment

is pretty low. However, his passive solar home has saved him approximately $18,000 in heating bills during this same period. Savings on electricity from the PV and savings on heating bills resulting from passive solar heating have more than paid for his PV system.

Dan accrued additional savings by avoiding the line extension fee of $2,000. Avoiding the $20/month meter-reading fee the local utility wanted to charge saves him another $240 per year or nearly $3,120 in the first 13 years. Even though he had to replace his batteries after 11 years, which cost $3,400, Dan's clearly ahead. What's the balance?

System cost, including the new batteries: $20,400.

Systems savings: $27,120.

Economics is where the rubber meets the road. Comparing solar electric systems against the "competition," calculating the return on investment, or comparing strategies using discounting and net present value, gives a potential buyer a much more realistic view of the feasibility of solar energy at a particular site. Just don't forget to think about all the opportunities to save money. If you invest in efficiency measures to lower the system cost, remember that those efficiency measures will provide a lifetime of savings, helping to underwrite your PV system. As we pointed out in Chapter 1, and have mentioned in this chapter, economics is not the only metric by which we base our decisions. Energy independence, environmental values, reliability, the cool factor, bragging rights, the fun value, and other factors all play prominently in our decisions to invest in renewable energy.

People often invest in renewable energy because they want to do the right thing. If you want to invest $12,000 to $60,000 in a PV system to lower your carbon footprint, give you peace of mind, or to live by your values, do it. It's *your* life. It's *your* money.

CHAPTER 5

SOLAR ELECTRIC SYSTEMS — WHAT ARE YOUR OPTIONS?

As discussed in Chapter 4, PV systems fall into three categories: (1) grid-connected, (2) grid-connected with battery backup, and (3) off-grid. In this chapter, we'll examine each system and their main components. We'll discuss the pros and cons of each system and then explore hybrid renewable energy systems — those that couple PV electric systems with other renewable energy technologies such as wind energy. The information in this chapter will help you decide which system suits your needs, lifestyle and pocketbook.

Grid-connected PV Systems

Grid-connected PV systems are the most popular solar electric system on the market today. As shown in Figure 5-1, grid-connected systems are so named because they are connected directly to the electrical grid — the vast network of electric wires that spans the nation and crisscrosses your neighborhood. These systems are also sometimes referred to as *batteryless grid-connected* or *batteryless utility-tied* systems because they do not employ batteries to store surplus electricity. Technically, the term utility-tied is more accurate. The term "grid" refers to the high-voltage electrical transmission system, not the local electrical distribution systems in cities and towns. Customers are actually connected to the grid through their local utility's distribution system, so the terms "utility-connected" or "utility-tied" are more accurate than "grid-connected" and "grid-tied," which are the terms commonly used by the renewable energy community. But, since it is used most often, we'll stick with the term "grid-connected."

In grid-connected systems, the local electrical distribution system (the wires that run by your house) accepts surplus electricity — electricity produced by a PV system in excess of the demand of the homeowner or business. When the PV system is not generating electricity, for example, at night, the utility supplies electricity to the home or business. In a sense, then, the utility grid serves as a storage medium.

As shown in Figure 5-1, a grid-connected system consists of five main components: (1) a PV array, (2) an inverter designed specifically for grid connection, (3) the main service panel or breaker box, (4) safety disconnects, and (5) meters.

To understand how a batteryless grid-connected system works, let's begin with the PV array. The PV array produces DC electricity. It flows through wires to the *inverter*, which converts the DC electricity to AC electricity. (For a description of AC and DC electricity, see the sidebar "AC vs. DC Electricity.")

The inverter doesn't just convert the DC electricity to AC, it converts it to grid-compatible AC — that is, 60-cycles-per-second, 120-volt (or 240-volt) electricity. (See sidebar "Frequency and Voltage" for more on this.) Because the inverter produces electricity in sync with the grid, inverters in these systems are often referred to as *synchronous inverters*.

The 120-volt or 240-volt AC produced by the inverter flows to the main service panel, a.k.a. the breaker box. From here, it flows to active loads — that is, to electrical devices that are operating. If the PV system is producing more electricity than is needed to meet these demands — which is often the

AC vs. DC Electricity

Electricity comes in two basic forms: *direct current* and *alternating current.* Direct current electricity consists of electrons that flow only in one direction through an electrical circuit. It's the kind of electricity produced by a flashlight battery or the batteries in cell phones, laptop computers, or the electrical systems in our automobiles. It is also the kind of electricity produced by photovoltaic modules.

Most other sources of electricity, including wind turbines and conventional power plants, produce alternating current electricity. Like DC electricity, AC electricity consists of the flow of electrons through a circuit. However, in alternating current, the electrons flow back and forth. That is, they change (alternate) direction in very rapid succession, hence the name "alternating current." Each change in the direction of flow (from left to right and back again) is called a cycle.

In North America, electric utilities produce electricity that cycles back and forth 60 times per second. It's referred to as 60-cycle-per-second — or 60 Hertz (Hz) — AC. The hertz unit commemorates Heinrich Hertz, the German physicist whose research on electromagnetic radiation served as a foundation for radio, television, and wireless transmission. In Europe and Asia, utilities produce 50-cycle-per-second AC.

A Inverter
B Breaker box
   (main service panel)
C Utility meter
D Wire to utility line
E Circuits to household
   loads

a

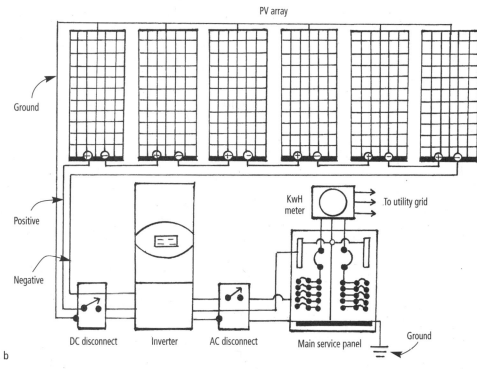

b

Fig. 5-1 a and b:
Schematic of
Grid-connected
PV System.
(a) overview and
(b) drawing of
system wiring
and system
components.

case on sunny days — the excess automatically flows onto the grid, or, as the experts say, the surplus is "back fed" onto the grid.

As shown in Figure 5-1, surplus electricity travels from the main service panel through the utility's electric meter, which is typically mounted on the outside of the house. It then flows through the wires that connect the home or business to the utility lines. From here, it travels along the power lines running by your home or business where it is consumed in neighboring homes and businesses. Once the electricity is fed onto the grid, the utility treats it as if it were theirs. End users pay the utility directly for the electricity you generated.

In most locations, an electric meter monitors the contribution of small-scale producers to the grid. The meter also keeps track of electricity the utility supplies to these homes or businesses when their PV systems aren't producing enough to meet their demands or when the PV system is not operating, for example, at night. (To learn how an electric utility measures a producer's output and how they "pay" for it, check out the accompanying sidebar, "Net Metering in Grid-connected Systems.")

In addition to the utility electric meter — or meters (some utilities require two or more meters) — that monitor the flow of electricity to and from the local utility grid, code-compliant grid-connected solar electric systems also contain two safety disconnects. Safety disconnects are manually operated switches that enable service personnel to disconnect key points in the system to prevent electrical shock when servicing the system.

As shown in Figure 5-1, the first disconnect is located between the solar array and the inverter. This is a DC disconnect. The DC disconnect may be separate or be built into the inverter. The manual disconnect allows

Frequency and Voltage

Alternating current is characterized by a number of parameters, two of the most important being frequency and voltage. Frequency refers to the number of times electrons change direction every second and is measured as cycles per second. (One cycle occurs when the electrons switch from flowing to the right then to the left then back to the right again.) In North America, the frequency of electricity on the power grid is 60 cycles per second (also known as 60 Hertz).

The flow of electrons through an electrical wire is created by a force. Scientists refer to this mysterious electromotive force as *voltage*. The unit of measurement for voltage is *volts*.

Voltage is a more difficult electrical term to understand. You can think of it as electrical pressure, as it is the driving force that causes electrons to move through a conductor such as a wire; without this force, electrons will not move through a wire. Voltage is produced by batteries in flashlights, solar electric modules, wind generators, and conventional power plants.

the operator to *terminate* the flow of DC electricity from the array to the inverter in case the inverter needs to be serviced. That is, the DC disconnect will *isolate* the inverter from the array. (See sidebar for a safety tip.)

Although not shown in Figure 5-1, grid-connected systems also contain an AC disconnect, which is often simply a breaker located in the breaker box or main service panel. The AC disconnect is used to terminate the flow of AC electricity from the grid to the inverter, and vice versa.

Another AC disconnect switch is often required by the local utility. Shown in Figure 5-1, this disconnect must be mounted outside the home or business, typically in close proximity to the building's utility service entrance (where electrical wires enter a building). It must be readily accessible to utility workers. This disconnect must also be lockable. That is, it must contain a switch that can be locked in the open position by utility workers so no electricity flows to or from the grid. This disconnect is required so workers can isolate PV systems from the electrical grid and work on electrical lines without fear of shock if, for example, a line in your area goes down in an ice storm.

The ability to lock an AC disconnect provides assurance that a family member can't accidentally turn the system back on before utility workers have completed their repairs. Utility workers use their own locks. AC disconnects must also be clearly labeled "Interconnection Disconnect Switch."

AC disconnects must also be visible. That doesn't mean they need to be in a position for

Safety Tip

DC disconnects should only be shut off after first disconnecting the AC disconnect. The AC disconnect is often a circuit breaker in a home's load center (that is, the breaker box or main service panel). The AC disconnect terminates the flow of current from the utility, shutting down the inverter. Once the AC disconnect is switched off, the DC disconnect can then be opened (switched off). This completely isolates the inverter from all voltage sources.

easy viewing — although they do need to be in plain sight. Rather, "visible" refers to the fact that an AC disconnect must contain a switch that allows the utility worker to see that the electrical circuit has physically been broken. A circuit breaker won't do, because the internal workings of the breaker can't be seen.

For many years, lockable AC disconnects were considered critical for the safety of utility personnel. Both 120- and 240-volt electricity from an inverter increase to thousands of volts on the utility distribution line after it passes through the distribution transformer. (That's the transformer on the pole or pad outside your home. It boosts the voltage of electricity back fed onto the grid from 120 volts to about 12,000 volts.)

Although utility-company-accessible, lockable, visible AC disconnects are required by many utilities, large California and Colorado utilities with thousands of solar- and wind-electric systems now online have recently dropped this requirement. They've found that AC disconnects are not needed because grid-compatible (synchronous) inverters automatically shut off when the utility power goes down. Properly installed PV systems will not back feed onto a dead grid. Period.

As we'll discuss in Chapter 6, grid-compatible inverters monitor line voltage and frequency (the frequency and voltage of electricity on the grid). When they detect a change in either, for instance, a drop in voltage due to a power outage, the inverter automatically shuts down — and stays off until power is restored, and then waits for a short period, typically five minutes, before resuming normal operation. As a result, no electricity can flow onto the grid when grid power is not present.

Lockable disconnects are redundant because UL-listed inverters meet interconnection safety standards and can be relied upon to disconnect if the grid goes down. Although the automatic disconnect feature works reliably, when it comes to safety a second line of defense is considered prudent by some. The second line of defense is the lockable disconnect switch just described. When human life is at stake, prudence dictates that you not rely on only one system for safety. The lockable disconnect is a backup to the safety features of the inverter. It is also on the outside of the building, where the utility worker can see it — and see that it is open (disconnected). Utility workers can also put their own locks on them to be sure that it stays open (disconnected). The inverter is inside the building where the utility worker cannot see it. How could a utility worker know there is a UL-listed inverter inside? Maybe the homeowner has cobbled together something that feeds directly onto the grid, with no safety functions. Every source of power in a utility's system, every generator and every transformer, has a visible disconnect switch. Since the utilities have backup safety disconnects for all their own equipment, some assert that it is reasonable for them to require renewable energy system owners to have them too?

The Pros and Cons of Grid-connected Systems

Batteryless grid-connected systems represent the majority of all new solar electric systems in the United States. Although popular, they do have their pluses and minuses, summarized in Table 5-1.

Table 5.1 Pros and Cons of Batteryless Grid-Tied Systems	
Pros	**Cons**
Simpler than other systems	Vulnerable to grid failure unless a backup generator or an uninterruptible power supply is installed
Less expensive	
Less maintenance	
Unlimited supply of electricity (unless the grid is down)	
More efficient than battery-based systems	
Unlimited storage of surplus electricity (unless the grid is down)	
Greener than battery-based systems	

Metering and Billing in Grid-connected Systems

The idea of selling electricity to a local utility appeals to many people. But how do utilities keep track of the two-way flow of electricity from grid-connected renewable energy systems so they know how much you supply to them and how much they sell to you?

In many cases, utilities use the existing dial-type electric meter. This meter measures kilowatt-hours of energy and can run forward or backward, i.e., it is bidirectional. It runs forward (counting up) when your home or business is drawing electricity from the electrical grid — for instance, if a home or business is using more energy than its PV system is producing. It runs backward (counting down) when the PV system is producing more power than is being used.

Some utilities install two dial-type meters, one for each direction — that is, one to tally the electricity delivered to the grid and another to keep track of electricity supplied by the grid. These meters are modified with a ratchet mechanism so they only turn — and register kilowatt-hours — in one direction. (Remember that utilities typically charge to install the second meter; they also charge a separate monthly fee to read a second meter!)[1]

In many new installations, utility companies install digital net meters that tally electricity delivered to and supplied by a home or business. They keep separate totals of the electricity coming from and going to the grid.

If a bidirectional meter is used, the utility typically offers a billing service called *net metering*. If two meters, or a digital meter that keeps two totals, are installed, the arrangement is referred to as *net billing* or *"buy-sell."*

Net metering is an arrangement that gives customers an incentive to feed their surplus renewable energy onto the grid. Each month in which customers have surplus energy, the surplus is credited to their account. (Surpluses are typically credited in kilowatt-hours, not dollars.) In a month with surplus energy, customers are not billed for any energy, nor are they paid for the surplus. They are only billed for the monthly service charge. Excess electricity rolls over to the next month, like minutes on a cell phone. In a subsequent month during which the customers' consumption is greater than their production, their kilowatt-hour account is debited to cover the net consumption. If the account balance is adequate to cover the net consumption, the customer is not billed for energy. If the account balance is not adequate to cover the net consumption, the customer is billed only for the remainder. The account is never allowed to go negative.

The "net" in net metering refers to the fact that the customer is billed only for net consumption, i.e., what remains after credits for surpluses are applied. Usually the account is reconciled once a year. This billing arrangement is known as *annual net metering*. Annual net metering means that utilities carry the surplus kilowatt-hours from one month to the next up to a year, just like many cell phone companies carry forward surplus minutes of call time. Any kilowatt-hours remaining in the account at the end of the year are either transferred to the utility, purchased at the retail price of electricity, i.e., the same price that the ☞

customer pays the utility, or purchased by the utility at their wholesale rate. In the least desirable arrangement for the customer, the surplus is simply forfeited to the utility. In each case, however, the account balance is set to zero and the net metering arrangement starts over for the following year.

The advantage of annual net metering is that it accommodates the seasonal variation in a PV system's production. In the summer months in many climates, PV systems produce more electricity than is consumed by homes or businesses. These surpluses can be "banked" with the utility.[2] In the winter months, when PV systems typically produce less electricity than is consumed, credits are withdrawn from the "bank" and applied to the bill. In the most favorable arrangement, the credits are applied to the bill at the retail rate.

To see how this works, consider an example. Suppose that a customer with a PV system mounted on her garage delivered 800 kWh of electricity to the grid during the month of August, a fairly sunny month in most locations in North America. Suppose also that the customer consumed 600 kWh from the grid during the month. For this month, the customer would be credited with the net production of 200 kWh and would not be billed for any electricity, although she would be billed for the normal customer service charge of $5 to $20. Utilities may also charge taxes and other incidentals, for example, surcharges to fund pollution abatement or renewable energy rebate programs. In this example, then, the customer would simply pay the customer service charge and applicable surcharges.

Let's assume that the next month, September, was unusually cloudy, so the customer delivered only 400 kWh of electricity, but the home consumed 550 kWh of electrical energy from the grid. In this month, the utility would deduct 150 kWh (the net consumption) from the customer's account, leaving a balance of 50 kWh. Since the balance in the account was sufficient to cover the net consumption, the customer would not be billed for any energy, only the monthly service charge.

Now consider the next month of net metering, October. For this month, the customer produces 600 kWh of electricity and consumes 800. The net consumption is 200 kWh, but this exceeds the balance in the account. The utility would deduct 50 kWh from the account (reducing the balance to zero) and bill the customer for the remaining 150 kWh.

In some cases utilities reconcile the customer's electric bills each month. This arrangement is known as *monthly net metering.* To see how this works, let's consider an example.

Suppose that the customer produced a surplus of 200 kWh of electricity in the month of August. (That is, she delivered 200 more kWh of electricity than she withdrew from the grid. This is known as *net excess generation.*) Let's also suppose that the utility charges 14 cents per kilowatt-hour for electricity supplied by the grid and credits customers the same amount for surpluses they deliver to the system. In this instance, then, the customer would receive a check for $28 minus meter reading and other fees. (That's 200 kWh net excess generation x 14 cents per kWh minus meter reading and other fees.) ☞

If you're thinking that this could be a profitable venture, don't get your hopes up. We've only found two states — Minnesota and Wisconsin — that pay customers retail rates for monthly net excess generation. (Wisconsin only pays if the monthly net energy generation exceeds $25; otherwise the surplus is carried forward to the next month.) Most other utilities pay for surpluses at the avoided cost, that is, the cost of generating new power. (Most of these states reconcile monthly; nine reconcile annually.) Some states like Arkansas simply "take" the surplus without payment to the customer — it all depends on state law. (If you're not happy with your state law, you might want to consider working to change it!)

As a rule, monthly net metering is the least desirable option, especially if surpluses are "donated" to the utility company or reimbursed at wholesale rates. Annual reconciliation is a much better deal. It permits summertime surpluses, if any, to be "banked" to offset wintertime shortfalls, if any. However, don't forget that even with annual net metering, the end-of-the-year surplus, if any, may be lost, that is, forfeited to the utility.

The ideal arrangement, from a customer's standpoint, notes *Home Power*'s Ian Woofenden, is "to negotiate a continual roll-over, so there's no concern about losing credit for your solar-powered kilowatt-hours."

Net metering is mandatory in many states, thanks to the hard work of renewable energy activists and forward-thinking legislators. At this writing (June 2009), 41 states and the District of Columbia have implemented statewide net metering programs, although there are substantial differences among the states. Differences include who is eligible, which utilities are required to participate in the program, the size and types of systems that qualify, payment structure, and so on. Many states only require "investor-owned" utilities to offer net metering. In many states, municipal (city-owned) utilities and rural cooperatives are exempt from net metering laws.

Utilities that don't offer net metering may use a buy-sell or net billing system. In buy-sell schemes, utilities typically install two meters, one to track electricity the utility sells to the customer, and another to track electricity fed onto the grid by the customer. The problem with net billing arrangements is that, unlike net metering, utilities typically charge their customers retail rates for electricity used by the customer but pay customers wholesale rates for electricity fed onto the grid by a renewable energy system. For example, a utility may charge 10 to 15 cents per kilowatt-hour for electricity they supply to the customers, but pay only wholesale rates of 2 to 3 cents per kilowatt-hour for energy that customers deliver to the grid. How does this work out financially for the small-scale producer?

As you might suspect, not very well.

Suppose that a customer consumed 500 kWh of electricity from the grid but fed 1,000 kWh of electricity onto the grid in July. If the retail rate for electricity was 14 cents per kilowatt-hour and the wholesale rate was 3 cents per kilowatt-hour, the utility would charge the customer 14 cents per ☞

kilowatt-hour for electricity delivered to them, or $70 for the 500 kWh (500 kWh x 14 cents per kWh = $70). The utility would credit the customer 3 cents per kilowatt-hour for the 1,000 kWh of electricity delivered to the grid. That is, the customer would be paid or credited $30 for the 1,000 kWh of electricity sold to the utility (1,000 kWh x 3 cents per kWh = $30). As a result, the customer would end up owing the utility $40 ($70 - $30 = $40) plus a meter reading fee of $10 to $20.

Had the customer been able to sign an annual net metering contract, he or she would not have paid anything for electricity that month, and would have banked the 500 kWh excess production for use in later months, when it would offset excess consumption at the retail rate.

Although (at this writing) eight states do not have net metering laws, progress in this area is moving quickly. Utilities are finding that PV systems help shave peak load — that is, reduce electrical demand during periods of high use. As you can see in Figure 5.2, peak demand runs from about 11 AM to 6 PM. To meet electrical demand during this time, many utilities buy power from other utilities or fire up additional generating capacity, such as natural gas-powered electric plants. Both options are costly, very costly, so anything utilities can do to shave peak demand usually saves them money. Your surplus solar electricity actually helps the utility reduce costs and can save the company and other ratepayers money because it is produced during the peak demand hours. ■

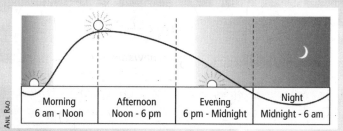

Fig. 5-2: *Peak Load (peak electrical demand) begins late in the morning and continues until early evening. PV systems can help shave peak load, that is, reduce the demand for electricity. This, in turn, makes it easier for utilities to supply inexpensive electricity to their customers, as explained in the text.*

| Morning 6 am - Noon | Afternoon Noon - 6 pm | Evening 6 pm - Midnight | Night Midnight - 6 am |

ANIL RAO

For More on Net Metering

The American Wind Energy Association has a great fact sheet on net metering at www.awea.org. Check out "Fact Sheets" under "Publications." For rules in your state, go to the Database for State Incentives on Renewables and Efficiency (dsireusa.org/). Click on the state, then click on net metering.

On the positive side, batteryless grid-connected systems are relatively simple. And, as such, they're generally less expensive than other options. Batteryless grid-tied systems are often 40% cheaper than off-grid systems and about 20% cheaper than grid-connected systems with battery backup. Because batteryless grid-connected systems contain no batteries, they require less maintenance than other systems.

Operators won't have to manage and maintain a costly battery bank, a topic discussed in Chapter 7. Nor will the owner have to replace costly batteries every five to ten years. Moreover, no batteries means no battery box or battery room, which can be costly to add to an existing home or incorporate into a new home. All of these factors add up to significant savings.

Another advantage of batteryless grid-connected systems is that they can store months of excess production. Although they don't literally store excess electricity on site, like a battery-based system, they "store" surplus electricity on the grid in the form of a credit on your utility bill (in areas where net metering is required by law). At night or when a PV system isn't producing enough electricity to meet demand, electricity can be drawn from the grid, using up the credit. The grid therefore serves as an unlimited battery bank. (For more on grid storage, see the accompanying sidebar, "Grid Storage — Is Electricity Stored on the Grid?")

Unlike a battery bank, you can never "fill up" the grid. It will accept as much electricity as you can feed it. In contrast, when batteries are full, they're full. They can't take on more electricity. As a result, surpluses generated in an off-grid PV system are typically lost. (The PV array is temporarily disconnected from the battery bank so no current flows to them when the batteries are full.)

Another advantage of grid-connected systems is that the grid is always there — well, almost always. Even if the Sun doesn't shine for a week, or two weeks, you have a reliable supply

Grid Storage — Is Electricity Stored on the Grid?

Many people are confused when renewable energy experts talk about storing electricity on the grid. There's good reason for this.

In a grid-connected system, surplus electricity flows onto the grid. It is not physically or chemically stored on the grid, however. It is consumed by one's nearest neighbors.

That said, electricity fed onto the grid is "effectively" stored thanks to net metering. When surplus electricity is back fed onto the grid, a homeowner or business is credited for the surplus. The utility banks the credit. Although it sells the electricity to another customer, the utility keeps track of the amount of electricity a PV customer has delivered to the grid, so it can credit the supplier later.

By keeping track of the surplus electricity produced by net-metered renewable energy system, the utility says, "You've supplied us with x number of kWh of electricity. When you need electricity, for example, when the Sun is down, we'll supply you with an equal amount at no cost." In a sense, the utility has stored the electricity for its customer. Bear in mind, however, that when the utility gives back the electricity you've "stored" on the grid, they supply you with electricity most likely generated by coal or nuclear energy or natural gas. The net effect on your utility bill is the same, however.

of electricity from the grid. With an off-grid system, at some point, the batteries run down. To meet your needs, you'll need to run your backup generator — or sit in the dark.

Yet another advantage of grid-connected systems with net metering is that utility customers suffer no losses when they store surplus electricity on the grid. With batteries, they do. As described in detail in Chapter 7, when electricity is stored in a battery, it is converted to chemical energy. When electricity is needed, the chemical energy is converted back to electrical energy. As much as 20 to 30% of the electrical energy fed into a battery bank is lost due to conversion inefficiencies.

In sharp contrast, electricity stored on the grid comes back in full. If you deliver 100 kWh of electricity, you can draw off 100 kWh. (This assumes annual net metering with the year-end reconciliation at retail price.) We'd be remiss if we didn't point out the grid has losses too, but net metered customers get 100% return on their stored electricity.

Another notable advantage of batteryless grid-tied systems is that they are greener than battery-based systems. Although utilities aren't the greenest businesses in the world, they are arguably greener than battery-based systems, because batteries require an enormous amount of energy to produce. Lead must be mined and refined. Batteries must be assembled and shipped to homes and businesses, requiring energy. Batteries contain toxic sulfuric acid and lead. Although old lead-acid batteries are recycled, they're often recycled under less than ideal conditions. (There's at least one US company [Deka]

that recycles its own batteries in the States.) Some companies ship batteries to less developed countries, where children are often employed to remove the lead plates along the banks of rivers. Acid from batteries may contaminate surface waters.

Yet another advantage of batteryless grid-connected systems, when net metered, is that they can provide an economic benefit. In sunny sites, these systems may produce surpluses month after month. If the local utility net meters and pays for surpluses at the end of the month or end of the year at retail rates, these surpluses can generate income that helps reduce the cost of PV systems and the annual cost of producing electricity.

On the downside, grid-connected systems may require careful and sometimes annoyingly difficult negotiations with local utilities. While many utilities are cooperative, others throw up roadblocks. Some utilities, especially some of the rural electric associations, are hostile to the idea of buying electricity from customers. (We'll discuss this topic in detail in Chapter 9.)

Another downside of batteryless grid-connected PV systems, worth serious consideration, is that they are vulnerable to grid failure. That is, when the grid goes down, so does the PV system. A home or business cannot use the output of a batteryless photovoltaic system when the grid is not operational. Even if the Sun is shining, batteryless grid-tied PV systems shut down if the grid experiences a problem — for instance, if a line breaks in an ice storm or lightning strikes a transformer two miles from your home or business, resulting in a power outage. Even though the Sun is shining, you'll get no power from your system.

If power outages are a recurring problem in your area and you want to avoid service disruptions, you may want to consider installing a standby generator that switches on automatically when grid power goes down. Bear in mind, however, that a standby or backup generator takes many seconds to start and come on line, so your power is interrupted during this time.

If you want to avoid this temporary interruption, you could install an uninterruptible power supply (UPS) on critical equipment such as computers. A UPS has a battery pack and an inverter. If the utility power goes out, the UPS will supply uninterrupted power until its battery gets low.[3] Or, as discussed in the next section, you may want to consider installing a grid-connected system with battery backup. In this case, batteries provide backup power to a home or business when the grid goes down.

Grid-connected Systems with Battery Backup

Grid-connected systems with battery backup are also known as *battery-based utility-tied* systems or *battery-based grid-connected* systems — take your pick. These systems ensure a continuous supply of electricity, even when an ice storm wipes out the electrical supply to you and the utility company's other 1.5 million customers in their service area.

As shown in Figure 5-3, a grid-connected system with battery backup contains all of the components found in grid-connected systems:

(1) a PV array, (2) an inverter, (3) safety disconnects, (4) a main service panel, and (5) meters to keep track of electricity delivered to and drawn from the grid. Although grid-connected systems with battery backup are similar to batteryless grid-connected systems, they differ in several notable ways.

One of the most important differences is the type of inverter. Inverters in battery-based grid-connected systems are very different from inverters in batteryless grid-tied systems. You can never change one into the other or use one in place of the other. (We'll describe the differences in more detail in Chapter 6.)

Another notable difference is that battery-based grid-connected systems require a battery bank. The third difference is a meter that allows the operator to monitor the flow of electricity into and out of the battery bank. The fourth difference is the charge controller.

Batteries for grid-connected systems with battery backup are either flooded lead-acid batteries or, more commonly, low-maintenance sealed lead-acid batteries. Because batteries are discussed in Chapter 7, we'll highlight only a few important considerations here.

The first point worth noting is that battery banks in grid-connected systems are typically small. That's because they are typically sized to provide sufficient storage to run a few critical loads for a day or two while the utility company restores electrical service. Critical loads might include a few lights, the refrigerator, a well pump, the blower of a furnace, or the pump in a gas- or oil-fired boiler. Those who want or need full power during outages must install much larger and costlier

A Charge controller
B Inverter
C Breaker box (main service panel)
D Utility meter
E Wire to utility service
F Circuits to household loads
G Back up battery bank

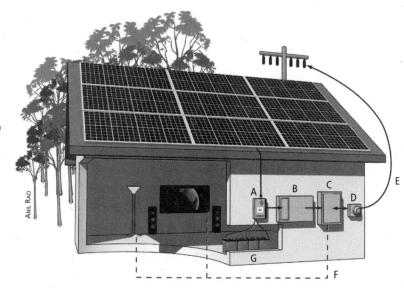

a

Fig. 5-3 a and b: *Grid-connected System with Battery Backup (a) overview and (b) drawing of system wiring and system components.*

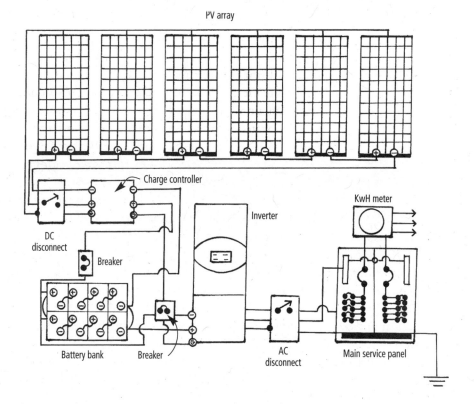

b

battery banks or install generators to meet their demand for electricity.

It is worth noting, too, that in battery-based grid-tied systems, batteries are called into duty only when the grid goes down. They're a backup source of power; they're not there to supply additional power, for example, to run loads that exceed the PV system's output. When demand exceeds supply, the grid makes up the difference, not the batteries. When the Sun is down, the grid, not the battery bank, becomes a home or business's power source.

It is also important to point out that battery banks in grid-connected systems are maintained at full charge — day in and day out — to ensure a ready supply of electricity should the grid go down. Keeping batteries fully charged is a high priority of these systems. Therefore, whenever power is being produced by the PV array, it first flows to the batteries. (Batteries require a small amount of power to "float" at a fully charged state, described in Chapter 7.) All of the power beyond what it takes to maintain this float charge is then sent to active loads in the home. When production exceeds consumption, the remaining electricity flows out, onto the grid. Unfortunately, some manufacturers have designed their inverters to maintain a float charge (a small flow of current) to the batteries using utility power at night or during periods when the system is not producing power. Other manufacturers' inverters allow the batteries to rest for short periods when the system is not producing, for example, overnight, and then resume floating the battery pack at

full charge when the Sun comes up the next day.

Maintaining a fully charged battery bank requires a fair amount of electricity over the long haul. That is to say, a small portion of the electricity a PV system generates and grid power are devoted to keeping batteries full at all times. This reduces system efficiency. The reduction in efficiency increases with the size of the battery bank, which is based on the size of the critical loads that require backup power. It also increases as the battery ages.

Batteries require a continual input because they self-discharge. That is, they lose electricity when sitting idly by. You've seen it happen to flashlight batteries or car batteries sitting idle for months. Because of this, batteries require a continual electrical charge and therefore become a regular load on a renewable energy system.

In the best case, topping off batteries will consume about 5 to 10% of a system's output. In the worst case — that is, in a system with a low-efficiency, unsophisticated inverter that is used to charge a large or older battery bank — it may approach 50%.

Battery banks in grid-connected systems don't require careful monitoring like those in off-grid systems, but it is a very good idea to keep a close eye on them. When an ice storm knocks out power to your home or business, the last thing you want to discover is that your battery bank quietly died on you last year. For this reason, grid-connected systems with battery backup typically include a meter to monitor the total amount of electricity stored in the battery bank. These meters give readings in amp-hours or kilowatt-hours. (See the sidebar, "Amps and Amp-Hours" for definitions.) You'll learn more about meters to monitor batteries in Chapter 7.

Meters in battery-based systems also typically display battery voltage. For experienced renewable energy operators, battery voltage provides a general approximation of the amount of energy in a battery. If, for instance, a battery is not being charged or discharged, the higher the voltage, the more energy it holds. The lower the voltage, the less energy it stores. Because charging a battery bank raises

voltage and discharging lowers voltage, you only get an accurate state-of-charge voltage when the batteries have been "at rest" for a couple of hours — that is, they have not been charged or discharged for a couple of hours.

Another component found in grid-connected systems with battery backup is the *charge controller*, shown in Figure 5-3. The charge controller regulates the flow of electricity into a battery bank — but only when there's a power outage. It does so by monitoring the voltage of a battery bank during power outages. When a charge controller

Amps and Amp-Hours

Electricity is the flow of electrons through a wire. Like water flowing through a hose, electricity flows through wires at varying rates. The rate of flow, or current, depends on the voltage (electromotive force).

The flow of electrons through a conductor is measured is amperes, or "amps" for short. (An ampere is a certain number of electrons passing by a point per second.)[4] The greater the amperage, the greater the number of electrons flowing through the circuit. In low-voltage systems, like the 12-volt system in your automobile, ten times more current (amps) will flow through a conductor compared to that which flows through the 120-volt circuits in a home — if the wattage is the same. Put another way, a 120-volt circuit produces the same amount of power (watts) as a 12-volt circuit with one-tenth the current. As a result, modern PV systems tend to be wired at higher operating voltages, which makes

the systems more efficient and less costly. (Reducing the current reduces line loss and requires smaller wires, saving money.)

To gain a better understanding of power produced or power consumed, we have to consider the flow of energy over a period of time. An *amp-hour* is the quantity of electrons flowing through a wire equal to one amp for a period of one hour. This term is also frequently used to define a battery's storage capacity. A flooded lead-acid battery, for example, might store 420 amp-hours of electricity. Because this is a pretty unfamiliar term, we recommend that amp-hours be converted to kilowatt-hours, a much more familiar term. (Individuals buy, and utility companies sell, kilowatt-hours.) To calculate kilowatt-hours from amp-hours, simply multiply the voltage of a battery by the amp-hours. A 6-volt, 420 amp-hour battery, for example, stores 2,520 watt-hours or about 2.5 kWh.

"sees" that the batteries are fully charged — they have reached the "full" voltage set point and hence are in danger of overcharging — it terminates the flow of electricity to the batteries. This prevents the batteries from overcharging, which can seriously damage the lead plates.

When the grid is operating normally, the inverter assumes the role of the charge controller. It monitors battery voltage like a charge controller and sends excess power to the grid. Excess power is that over what is required to hold a battery bank at proper float voltage. (The float voltage is a preset level that indicates the batteries are fully charged.)[5]

Charge control is essential to battery-based systems because overcharging batteries can permanently damage their lead plates, dramatically reducing battery life. Batteries also require protection from discharging too deeply, referred to as "over discharging." Like overcharging, over discharging damages the lead plates of batteries, dramatically reducing their life. To prevent over discharging, charge controllers contain a low-voltage disconnect (LVD), although they typically only "cover" DC loads — that is, circuits that draw DC electricity directly from the battery bank to supply DC loads (DC appliances and DC lights). To protect against over discharging by AC loads, most PV systems rely on a low-voltage disconnect located in the inverter. It shuts the inverter down if the battery voltage drops too low.

Low-voltage disconnects in the charge controller and inverter terminate the flow of electricity out of the battery bank when the amount of electricity stored in batteries falls to 20% of the battery's storage capacity. In battery-based grid-tied systems, deep discharge is a rare event. Over discharging typically only occurs during extended utility power outages — that is, when the utility power is down and the batteries are being used to supply critical loads for extended periods.

Modern charge controllers and inverters often contain a function called *maximum power point tracking* (MPPT). Discussed in detail in Chapter 7, maximum power point tracking circuitry optimizes the output of a PV array, thus ensuring the highest possible output at all times.

Most charge controllers on the market also contain a high voltage/low voltage DC conversion function. As you will learn in Chapter 7, this feature allows the array to be wired as high as 60 or 72 volts (nominal) and still charge a 12-, 24- or 48-volt battery bank. This cuts down on the expense of the "home-run" wire — the wire from the array to the balance of system (BOS) components — that is, the rest of the PV system — and also allows the array to be located further from the BOS.

Pros and Cons of Grid-connected Systems with Battery Backup

Grid-connected systems with battery backup protect homeowners and businesses from power outages. They enable them to continue to run critical loads — that is, select loads — during outages. They may, for instance, allow homeowners to run an energy-efficient refrigerator and energy-efficient lighting while neighbors grope around in the dark and while the food in their refrigerators begins to rot. These systems allow businesses to protect computers and other vital electronic equipment so they can continue operations while their competitors twiddle their thumbs and complain about financial losses.

Although battery backup may seem like a desirable feature, it does have some drawbacks. For one, grid-connected systems with battery backup cost more to install — about 30% more. The higher cost, of course, is due to the added components, especially the charge controller and the batteries. Flooded lead-acid batteries and sealed batteries used in these and other renewable energy systems are expensive.

Remember, too, that a battery bank needs a safe, comfortable home. If you are building a new home or office, you need to add a well-ventilated battery box or battery room that stays warm in the winter and cool in the summer to house batteries.

In addition, unless you install a very large battery bank, your home or business won't be protected against extended blackouts — those lasting more than three or four cloudy days. Nor will you be able to meet all of the power requirements of your home or business.

Flooded lead-acid batteries also require periodic maintenance and replacement. As explained more fully in Chapter 7, to maintain batteries for long life, you'll need to monitor fluid levels regularly and fill batteries with distilled water every few months. Battery banks also need to be vented to the outside to prevent potentially dangerous hydrogen

gas buildup, which can lead to explosions and fires if ignited by a spark. In addition, battery banks need to be replaced periodically, whether you use them or not. Typical batteries used in this system require replacement every five to ten years, at a cost of two to three thousand dollars each time.

Yet another problem, noted earlier, is that battery-based grid-tied systems require a substantial portion of the daily renewable energy production just to keep the batteries topped off (fully charged).

Because grid-connected systems with battery backup are expensive and infrequently required, few people install them. For about ten years, Mick Sagrillo, a wind expert who generates his electricity via a couple of wind turbines, had a huge battery bank in the cellar of his home in Wisconsin. This large and costly battery bank was called into service only a couple of times because power outages in his area are infrequent and short-lived, rarely lasting longer than 10 to 20 minutes. In ten years, his family only experienced a few power outages that lasted more than a few hours. In a case of out of sight, out of mind, Mick freely didn't perform the required maintenance, so when he needed them, the batteries didn't perform well. They'd lost much of their storage capacity and thus provided only a fraction of the electricity his family needed. Because of his experience, he rarely recommends battery backup to clients.

When contemplating a battery-based grid-tied system, you need to ask yourself three questions: (1) How frequently does the grid fail in your area? (2) What critical loads are present and how important is it to keep them running? (3) How do you react when the grid fails?

If the local grid is extremely reliable, you don't have medical support equipment to run, your computers aren't needed for business or financial transactions, and you don't mind using candles on the rare occasions when the grid goes down, why buy, maintain and replace costly batteries?

In some cases, people are willing to pay for the reliability that a battery bank brings to a grid-connected system. One of Dan's clients in British Columbia buys and sells stocks, bonds and currency for huge accounts. He can't experience downtime during active trading — not for a second. As a result, he opted for a grid-connected system with battery backup for his home and office. See Table 5-2 for a quick summary of the pros and cons of battery-based grid-connected systems.

Off-Grid (Stand-alone) Systems

Off-grid systems are designed for individuals and businesses that want to or must supply all of their needs via solar energy — or a combination of solar and wind or some other renewable source. As shown in Figure 5-4a, off-grid systems bear a remarkable resemblance to a grid-connected system with battery backup. There are some noteworthy differences, however. Of course, there *is* no grid connection. As you can see in Figure 5-4a, there are no power lines running from the house or business to the grid. These systems "stand alone."

As illustrated in Figure 5-4, the main source of electrical energy in an off-grid system is the

Table 5.2 Pros and Cons of a Battery-based Grid-Tied System	
Pros	Cons
Provide a reliable source of electricity	More costly than battery-less grid-connected systems
Provide emergency power during a utility outage	Less efficient than battery-less grid-connected systems
	Less environmentally friendly than batteryless systems
	Require more maintenance than batteryless grid-connected systems

PV array. Electricity flows from the PV array to the charge controller. The charge controller monitors battery voltage and delivers DC electricity to the battery bank. When electricity is needed in a home or business, it is drawn from the battery bank via the inverter. The inverter converts the DC electricity from the battery bank, typically 24 or 48 volts in a standard system, to higher-voltage AC, either 120 or 240 volts, which is required by households and businesses. AC electricity then flows to active circuits in the house via the main service panel.

Off-grid systems often require a little "assistance," in the form of a wind turbine, micro hydro turbine, or a gasoline or diesel generator, often referred to as a gen-set. One or more of these energy sources helps make up for shortfalls.

Although backup generators are commonly used in off-grid renewable energy systems,

A   Charge controller
B   Inverter
C   Breaker box
D   Circuits to household loads
E   Battery bank
F   Back up generator

a

Fig. 5-4 a and b:
Off-grid System.
a) overview and
(b) drawing of
system wiring and
system components.

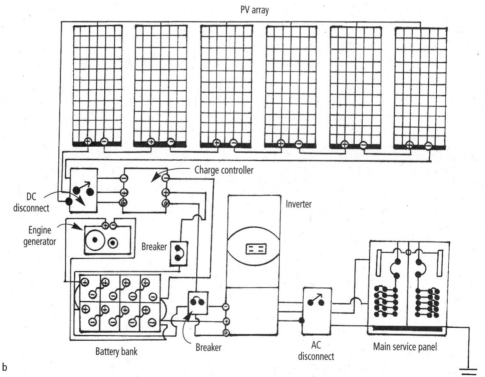

b

some experts, like Mick Sagrillo, contend that properly sized PV/wind hybrid systems rarely, if ever, require them. In fact, he's retrofitted numerous PV systems with wind generators to avoid the need for generator backup. Even though hybrid systems work well, the majority of off-grid systems include generators. It takes a well-balanced mix of solar and wind resources to avoid the need for a gen-set. "A gen-set also provides redundancy," notes National Renewable Energy Laboratory's wind energy expert Jim Green. Moreover, "if a critical component of a hybrid system goes down temporarily, the gen-set can fill in while repairs are made." Finally, gen-sets also play a key role in maintaining batteries, a subject discussed in Chapter 7.

Off-grid systems with gen-sets also require battery chargers. They convert the AC electricity produced by the generator into DC electricity that's then fed into the battery bank. In olden days, battery chargers were separate items, but they're now built into the inverter and operate automatically. When a generator is started and the inverter senses voltage at its input terminals, it transfers the home loads over to the generator through an internal, automatic transfer switch, and begins charging the battery from the generator.

Like grid-connected systems with battery backup, an off-grid system requires safety disconnects — to permit safe servicing. DC disconnects, with appropriately rated fuses or breakers, are located between the PV array and the charge controller, between the charge controller and the battery bank, and between the battery and the inverter.

These systems require charge controllers to protect the batteries from overcharging and low-voltage disconnects (LVDs) to prevent deep discharge of the battery bank. LVDs for DC loads are typically located in the charge controller, while LVDs for AC loads are housed in the inverter. Modern charge controllers also contain maximum power point tracking circuitry, mentioned earlier.

Charge controllers not only prevent batteries from being overcharged, they also prevent reverse current flow from the battery back to the array at night. Although reverse current flow is typically very small, it is best to be avoided. Modern-day charge controllers make this a non-issue.

As is evident by comparing schematics of the three types of systems, off-grid PV systems are the most complex. Moreover, some systems are partially wired for DC — that is, they contain DC circuits that are fed directly from the battery bank. DC circuits are used to service lights or DC appliances such as refrigerators or DC well or cistern pumps. (For a discussion of DC circuits and DC appliances, see the sidebar, "DC Circuits in an Otherwise AC World?")

To simplify installation of battery-based systems, many installers recommend use of a power center, such as the one shown in Figure 5-6. Power centers contain many of the essential components of a renewable energy system, including the inverter, the charge controller, and fused safety disconnects — all prewired. This makes an electrician's job easier. Power centers also provide busses (connection points) to which the wires leading to the

DC Circuits in an Otherwise AC World?

Most modern homes and businesses operate on alternating current electricity. However, off-grid homes supplied by wind or solar electricity — or a combination of the two — can be wired to operate partially or entirely on direct current electricity to power DC lights, refrigerators and even ceiling fans. Why wire a home or cottage for DC?

One reason is that DC systems do not require inverters. Electricity flows directly out of the battery bank to service loads. This, in turn, can reduce the cost of the system as household-sized inverters cost $1,000 to $4,000. However, cost savings created by avoiding an inverter may be offset by higher costs elsewhere. For example, DC appliances typically cost more than AC appliances —considerably more. DC ceiling fans, for instance, cost four times more than comparable AC models. You could pay $200-$250 for a DC model, but $40-$50 for a comparable AC ceiling fan.

DC appliances and electronics are not only more expensive, they are more difficult to find. You won't find them at national or local appliance and electronics retailers.

Many DC appliances are tiny, too. Most DC refrigerators, for example, are miniscule compared to the AC models used in homes. That's because DC appliances are primarily marketed to boat and recreational vehicle enthusiasts, and there's not a lot of room in a boat or recreational vehicle for large appliances.

For these and other reasons, DC-only systems are rare. They are typically installed only in remote cabins and cottages that are only occasionally used.

Another reason for avoiding use of an inverter is efficiency. As you will see in Chapter 8, most inverters for off-grid systems are about 90 to 95% efficient. That is, they consume some energy — 5 to 10% — when converting DC to AC and boosting the voltage. Bypassing the inverter with a DC circuit to the water pump or refrigerator reduces this loss. Over the long haul, bypassing the inverter can result in large savings.

Although DC circuits may seem like a good idea, *Home Power*'s Ian Woofenden argues in favor of caution when considering this approach. He notes that water pumping typically does not require a lot of electricity, unless, of course you are irrigating a lawn or large garden or watering livestock. Ceiling fans also are not that significant in the big picture. However, Ian argues that it's easier to make a case for a DC refrigerator or DC lighting, which are more substantial electrical loads in homes.

Even though modern refrigerators are much more energy efficient than their predecessors, they still are major energy consumers in modern homes. Refrigeration often accounts for 10 to 25% of a family's daily electrical demand. Because refrigerators consume so much electricity, a DC unit like those offered by Sun Frost or Sun Danzer, can save a substantial amount of energy over the long run (Figure 5-5). Those thinking about an off-grid system, like one of Dan's clients who runs a rustic "hotel" in Tulum, Mexico with no grid power nearby, may want to consider DC refrigerators and DC freezers like the Sun Danzer chest freezer.

While efficient, DC refrigerators and freezers do not come with the features that many Americans ☞

expect, such as automatic defrost. They also cost much more than standard or even high-efficiency AC units. Because of these reasons, we rarely recommend the inclusion of DC circuits for off-grid installations. They're only for a certain type of renewable energy user — people who are trying to wring every possible kilowatt-hour from a small system.

There are other reasons to think twice about DC circuits in an otherwise AC home. For one, low-voltage DC circuits must be wired with larger gauge wire. Richard Perez, founder of *Home Power* magazine, notes that electrical connections should be soldered, which adds cost. DC circuits also require special plugs and sockets that are more expensive than their AC counterparts.

Perez also notes that DC appliances are not typically as reliable or well built as their AC counterparts. They are, he says, primarily designed for intermittent use in recreational vehicles. DC appliances therefore wear out more quickly and thus require more frequent replacement.

As a case in point, you'll find that a DC blender costs about twice as much as an AC model. The blender, Perez jokes, has only two speeds — on and off. Moreover, it can only be ordered from specialty houses by mail or via the Internet. The DC blender he and his wife bought died after fewer than eight months of use. As if that's not enough to dissuade you from DC appliances, Perez points out that many appliances have no DC counterparts.

Arguing in favor of AC installation, Perez notes, "The main advantage of using AC appliances is standardization. The wiring is standard — inexpensive, conventional house wiring. The appliances

Fig. 5-5: *Sun Frost Refrigerator. Superefficient, but pricey, this refrigerator was the kingpin of refrigerators for many years. Now, thanks to new laws and government-sponsored competitions among appliance manufacturers, many new super efficient refrigerators are available at a much lower cost.*

are standard and are available with a wide variety of features." Furthermore, "The appliances are designed for regular use, and most are reliable and well built."

In closing, Perez notes that while "the main disadvantage of using AC appliances in off-grid systems is the cost of the inverter and the energy lost due ☞

to inversion inefficiency," modern inverters have an average operating efficiency that is similar to the amount of energy lost in low-voltage DC wiring, especially if the wire runs are long or are not terminated properly. In other words, in the final analysis, DC may not save any energy at all by bypassing the inverter.

If you are thinking about installing an off-grid system in a home or business, your best bet is an AC system. Even so, you may want to consider installing a few DC circuits. Dan's off-grid home, powered by solar electricity and wind, uses AC electricity almost entirely. However, when wiring the home, he included a DC circuit to power a DC pump that pumps water from his cistern to the house. He also included a DC circuit to power three DC ceiling fans; although he bought AC fans and runs AC electricity to them because of their lower cost. (Were he to do this again, he'd install a DC fridge.)

You might want to consider installing a DC circuit in the utility room — or wherever the inverter is located — just to power a DC compact fluorescent light bulb in case of emergency! That way, if the system goes down at night, and you need to find out why, you'll have some light. Be sure to have a spare DC light bulb on hand, too. ■

Fig. 5-6: *Power Center.* Power centers like the one shown here contain all the key components of a PV system, including one or more inverters, charge controllers, fuses and disconnects.

battery bank, the inverter and the PV array connect.

Although power centers may cost a bit more than buying all the components separately, they are easier and cheaper to install. And, if you're hiring a professional to install the system, they will save on labor costs.

Power centers make inspectors happy — very happy (and a happy inspector is a good thing!). While some electrical inspectors know a fair amount about renewable energy systems, many haven't had much experience in this area. When one of the less experienced inspectors shows up on a job site and encounters a renewable energy system equipped with a power center that's approved by Underwriter's Lab (UL), he or she can be assured that all of the vital components are present and accounted for. Inspections and approvals are a snap.

Pros and Cons of Off-Grid Systems.

Off-grid systems offer many benefits, including total emancipation from the electric utility (Table 5-3). They provide a high degree of energy independence that many people long for. You become your own utility, responsible for all of your energy production. In addition, if designed and operated correctly, they'll provide energy day in and day out for many years. Off-grid systems also provide freedom from occasional power failures.

Off-grid systems do have some downsides. As you might suspect, they are the most expensive of the three renewable energy system options. Battery banks, supplemental wind systems and generators add substantially to the cost — often 60% more. They also require more wiring. In addition, you will need space to house battery banks and generators. Dan had to build an insulated, soundproof ventilated shed to house his backup generator to appease neighbors who complained about the noise. Batteries also require periodic maintenance and replacement every five to ten years, depending on the quality of batteries you buy and how well you maintain and treat them.

Although cost is usually a major downside, there are times when off-grid systems cost the same or less than grid-connected systems — for example, if a home or business is located more than a few tenths of a mile from

Table 5.3
Pros and Cons of an off-grid System

Pros	Cons
Provide a reliable source of electricity	Generally the most costly solar electric systems
Provide freedom from utility grid	Less efficient than battery-less grid-connected systems
Can be cheaper to install than grid-connected systems if located more than 0.2 miles from grid	Require more maintenance than batteryless grid-connected systems (you take on all of the utility operation and maintenance jobs and costs)

the electric grid. Under such circumstances, it can cost more to run electric lines to a home than to install an off-grid system. (We discussed this in Chapter 4.)

When installing an off-grid system, remember that you become the local power company and your independence comes at a cost to you. Also, although you may be "independent" from the utility, you will need to buy a gen-set and fuel, both from large corporations. Gen-sets cost money to maintain and operate, and you will be dependent on your own ability to repair your power system when something fails. Independence also comes at a cost to the environment. Gen-sets produce air and noise pollution. Lead-acid batteries are far from environmentally benign. As noted earlier, although virtually all lead-acid batteries are recycled, battery production is responsible for considerable environmental degradation. Mining and refining the lead, as noted earlier, are fairly damaging. Thanks to

NAFTA and the global economy, lead production and battery recycling are being carried out in many poor countries with lax or nonexistent environmental policies. They are responsible for some of the most egregious pollution and health problems facing poorer nations across the globe, according to small wind energy expert Mick Sagrillo. So, think carefully before you decide to install an off-grid system. For more on this topic, see the sidebar, "To Stand Alone or Not to Stand Alone?"

Hybrid Systems

As you've just seen, you have three basic options when it comes to solar electric systems. Each of these systems can be designed to include additional renewable energy sources. The result is a hybrid renewable energy system (Figure 5-7).

Hybrid renewable energy systems are extremely popular among homeowners in rural areas. One of the most active segments of the residential wind energy market is the PV-only system owners who are expanding their

To Stand Alone or Not to Stand Alone?

Many people speak to Dan about going off grid by installing a PV or PV and wind system, even though their homes or businesses are currently grid tied. While this may sound like a glorious way to live your life, the off-grid option comes at a price. When Mick Sagrillo consults with individuals who want to go off grid for philosophical reasons, he cautions them to consider the ramifications of this decision, especially if they are philosophically opposed to utilities and concerned about the environmental impacts of grid-generated electricity. He does not at all condone how utilities operate, but from a purely environmental perspective, he thinks it is extremely difficult to justify a battery bank and gen-set system over grid connection.

We discussed the downsides of batteries earlier, but haven't discussed the generator. The electric grid delivers electricity generated by coal-fired power plants at about 30-33% efficiency and nuclear power plants at 20% efficiency. A Honda gen-set charging a battery bank operates at about 5% efficiency, Sagrillo contends. The emissions from that coal-fired plant are regulated by the EPA — more or less. Although there are also emission regulations for small engines (manufactured since September 1, 1997), the emissions per kilowatt-hour of electricity are far greater from the backyard generator than from the coal-fired power plant.[6] In fact, if everyone had a gen-set in their backyard to meet their electrical demands, we'd all suffocate. Visit Cali, or Beijing, or any number of developing countries where generators are used to produce electricity and you'll see, smell and choke on the result.

If you are thinking about going off grid, the responsible approach might be to install a hybrid system in which solar electricity is supplemented by some other clean, renewable energy technology such as wind or micro hydro.

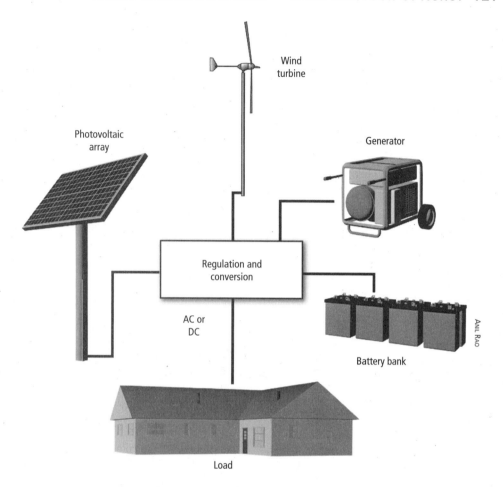

Wind turbine

Photovoltaic array

Generator

Regulation and conversion

AC or DC

Battery bank

ANIL RAO

Load

Fig. 5-7: Hybrid PV and Wind System

systems to incorporate wind energy, according to Mike Bergey of Bergey Windpower, a company that manufactures wind turbines.

Solar electricity and wind are a marriage made in heaven in many parts of the world. Why?

In most locations, solar energy and winds vary throughout the year. Solar radiation striking the Earth tends to be highest in the spring, summer and early fall. Winds tend to be strongest in the late fall, winter and early spring.

Table 5-4 shows that sunlight is relatively abundant in central Missouri from March through October. Table 5-5 shows that winds, however, pick up in October and blow through May. Together, solar electricity and wind can provide 100% of a family's or business's electrical energy needs. This complementary relationship is shown graphically in Figure 5-8.

In areas with sufficient solar and wind resources, a properly sized hybrid PV/wind

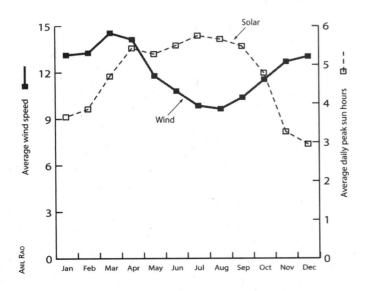

Fig. 5-8: *Graph showing Complementary Nature of Wind and Solar.*

system can not only provide 100% of your electricity, it may eliminate the need for a backup generator. Moreover, wind and PVs complement each another, you can install a smaller solar electric array and a smaller wind generator than if either were the sole source of electricity.

Hybrid systems may also make sense for those installing a grid-tied system with a buy-sell arrangement with a low sell price; a hybrid system reduces the amount of energy that must be bought (in net billing situations). For grid-tied systems with annual net metering or buy-sell with a sell price equal to or higher than the buy price, reducing seasonal variation in production has no advantages. The best strategy is to make as much electricity as

Table 5.4 Solar Resource near Gerald, Missouri measured in kWh/m² per day*													
Lat 38.325 Lon -91.297	Jan	Feb	Mar	Apr	May	Jun	Jul	Aug	Sep	Oct	Nov	Dec	Annual Average
Tilt 38	3.66	3.86	4.71	5.43	5.28	5.50	5.75	5.66	5.48	4.79	3.27	2.95	4.70

*Note: This data represents solar energy striking a collector mounted at an optimal angle for this location.

Table 5.5 Ten-Year Monthly Average Wind Speed Near Gerald, Missouri at 120 feet													
Lat 38.325 Lon -91.297	Jan	Feb	Mar	Apr	May	Jun	Jul	Aug	Sep	Oct	Nov	Dec	Annual Average
m/s	5.96	5.93	6.50	6.31	5.27	4.83	4.40	4.31	4.64	5.16	5.67	5.82	5.40
m/hr	13.33	13.26	14.54	14.12	11.79	10.8	9.84	9.64	10.38	11.54	12.68	13.02	12.08

m/hr

you can at the lowest cost. In this case, only one technology makes economic sense, the one with the lowest cost.

If the combined solar and wind resource is not sufficient throughout the year or the system is undersized, a hybrid system will require a backup generator — a gen-set to supply electricity during periods of low wind and low sunshine. Gen-sets are also used to maintain batteries in peak condition, discussed in Chapter 7, and permit use of a smaller battery bank.

Choosing a PV System

To sum things up, homeowners and businesses have three basic choices when it comes to installing a PV system. If they have access to the electric grid, they can install a battery-less grid-connected system, by far the cheapest and simplest option. Or they can install a grid-connected system with battery backup. If they don't have access to the grid, they can install an off-grid system. All of these systems can combine two or more sources of electricity, creating hybrid systems.

When Dan consults with clients who are thinking of building passive solar/solar electric homes, he usually recommends a grid-connected system for clients who live close to the utility grid. This configuration allows his clients to use the grid to store excess electricity. Grid-connected systems are cheaper than other options, too. Although they may encounter occasional power outages, as noted earlier, in most places, these are rare and transient occurrences — although many residents of Missouri and several other

Midwestern states went without power for 10 days in December 2007 after a crippling ice storm swept across the United States.

If you are installing a PV system on an existing home or business that is already connected to the grid, it is generally best to stay connected. Use the grid as your battery bank. When installing such a system, however, be sure to contact the local utility company and apprise them of your intentions. You'll need to sign an interconnection agreement with the company, and work out a billing agreement; hopefully, it will be for annual net metering that credits you at retail rates for surpluses.

Grid-connected systems with battery banks are suitable for those who want to stay connected to the grid, but also want to protect themselves from occasional blackouts. They'll cost more, but they provide peace of mind and real security. It's best to back up only the truly critical loads and make sure they are highly efficient. Doing so will reduce the size of the battery bank required to meet the loads, reduce the cost of the system, and improve the efficiency of the system. Some loads, like a forced air furnace, can be quite a challenge to back up. Not only is a furnace blower one of the larger loads in a home, it also runs during the time of the year with the least amount of sunlight.

Although more expensive than grid-connected systems, off-grid systems are often the system of choice for customers in remote rural locations. When building a new home in a rural location, grid connection can be pricey — so pricey that an off-grid system is less expensive. As noted in Chapter 4, you may be charged to run an electrical line to your home as well as for the meters and the utility disconnect. Hook-up fees can be upward of $50,000, if you live a half-mile from the closest electric lines. Rates vary, so be sure to check. Bob notes that a client in Central Indiana, who had a building site 7/8 of a mile from an existing power line, got a quote of $19,000 from his utility.

Some utility companies foot the bill for line extension, connection and metering. Be sure to check when considering which system you should install. "Utility policies vary considerably when it comes to line extension costs," notes NREL's Jim Green. "Sometimes, the utility absorbs much of the cost in the rate base. Others pass most or all of the cost to the new customer."

CHAPTER 6

UNDERSTANDING INVERTERS

The inverter is a modern marvel of electronic wizardry and an indispensable component of virtually all solar electric systems. Like batteries in off-grid systems, the inverter works day and night. Its main function is to convert DC electricity generated by a PV array into AC electricity, the type used in our homes and businesses. In battery-based systems, inverters contain circuitry to perform a host of additional useful functions.

In this chapter, we'll examine the role inverters play in renewable energy systems and then take a peek inside this remarkable device to see how it operates. We'll discuss all three types of inverters and discuss the features you should look for when purchasing an inverter. But the first question to answer is this: Do you need an inverter?

Do You Need an Inverter?

Although this may seem like a ridiculous question, it's not. Some solar applications operate solely on DC power and, as a result, don't require inverters. Included in this category are small solar electric systems like those in cabins, recreational vehicles, and sail boats that power a few DC circuits — for example, DC lights, refrigerators and pumps. It also includes direct water-pumping systems that produce DC electricity to power a DC water pump.

All other renewable energy systems require an inverter. The type of inverter one needs depends on the type of system. Grid-connected PV electric systems, the most common system, require a utility-compatible inverter. Off-grid systems require inverters that are designed to operate with batteries. Grid-connected systems with battery backup require an inverter that's grid- *and* battery-compatible.

How Does an Inverter Create AC Electricity?

To simplify the discussion, we'll start with inverters installed in off-grid systems, known as *battery-based inverters.*

Battery-based inverters perform two essential functions: (1) they convert DC to AC electricity, and (2) they increase the voltage. In off-grid systems, electricity from the PV array is first fed into a charge controller. It then flows into the batteries. When electricity

Fig. 6-1 a, b and c: *How an Inverter Works.*

is needed, DC electricity stored in the batteries flows to the inverter. The inverter converts the DC electricity into AC electricity.

As noted in Chapter 5, batteries are wired so they receive, store and supply 12-, 24- or 48-volt electricity. The most common wiring configurations are 24 and 48 volts. In a 24-volt battery-based system, the inverter converts 24-volt DC from the battery bank to 120-volt or 240-volt AC electricity — either 50 or 60 Hz, depending on the region of the world you're in. (In North America, AC is 60 Hz; in Europe and Asia, it's 50 Hz.)

The DC-to-AC conversion and the increase in voltage are performed by two separate, but integrated, components in an inverter. The conversion of DC to AC occurs in an electronic circuit commonly referred to as an *H-bridge*, for reasons that will be clear shortly. (It is also known as a *high-powered oscillator* or *power bridge*.) The second process, the increase in voltage, is performed by a transformer inside the inverter.

The Conversion of DC to AC

As shown in Figure 6-1a, the H-bridge consists of two vertical legs connected by a horizontal section, forming the letter H. Each leg contains two transistor switches that control the flow of electricity through the H-Bridge.

To see how the H-Bridge works, take a look at Figure 6-1b. As illustrated, DC electricity from the battery bank flows into the inverter from the battery and down the left leg (the arrows show the direction in which electricity flows). With switches 1 and 4 closed, low-voltage DC electricity flows from

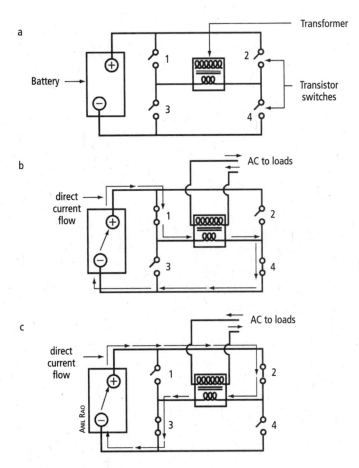

the battery into the vertical leg on the left side of the H-bridge then flows through the horizontal portion of the H-bridge and out the bottom of the right leg of the H-bridge. It then returns to the battery.

An electronic timer then flips the switches, opening 1 and 4 and closing 2 and 3. As shown in Figure 6-1c, electricity now flows from the battery into the vertical leg of the H-bridge on the right side of the diagram through switch 2. It then flows through the horizontal portion of the H-bridge (in the opposite direction) and out the vertical leg on the left side through switch 3. From here, it flows back to the battery.

Opening and closing the switches very rapidly allows DC electricity to flow first in one direction and then the other through the horizontal portion of the H-bridge and transformer. This alternating flow is AC electricity. A small controller (not included in the figures) regulates the opening and closing of the switches in the H-bridge. How's the voltage increased?

Increasing the Voltage

As illustrated in Figure 6-1, the horizontal portion of the H-bridge runs through one winding of a transformer. (The winding consists of a single wire, wrapped many times around a core.) A transformer is a rather simple electrical device that can either increase or decrease the voltage of AC electricity, as explained in the accompanying sidebar, "How Transformers Work."

The transformer in an inverter increases the 24-volt direct current flowing through the H-Bridge to 120-volt alternating current.

The label "AC to loads" in Figure 6-1b and c indicates the 120-volt AC electricity produced by the inverter.[1]

Square Wave, Modified Square Wave and Sine Wave Electricity

An inverter, such as the one we've been examining, produces alternating current electricity. However, as shown in Figure 6-2a, it is very choppy. This graph is a plot of the voltage of the electricity over time. To understand what this graph represents, let's step back a second and review what we know about alternating current electricity. As noted in Chapter 5, electrons flow back and forth in a wire carrying alternating current at a rate of 60 cycles per second. If you measured the voltage in a wire carrying AC electricity with an instrument that responds fast enough, you'd see that the current changes over time. It has to because the flow of electrons shifts direction — albeit very rapidly. When the electrons are flowing one way, say to the right, the voltage climbs very rapidly to +120 volts. When the flow of electrons shifts direction, the voltage very rapidly drops to -120 volts. (The positive and negative signs refer to the "polarity" of the voltage. That is, whether it is above or below a reference voltage of zero.) The graph in Figure 6-2a indicates the voltage over a very short period of time. The voltage shift from zero to +120 volts then back down to zero takes $1/120^{th}$ of a second (one half of $1/60^{th}$ of a second) for 60 Hz alternating current.

The basic H-bridge circuit found in some older inverters produces an almost perfect *square wave*, which is not a very usable form

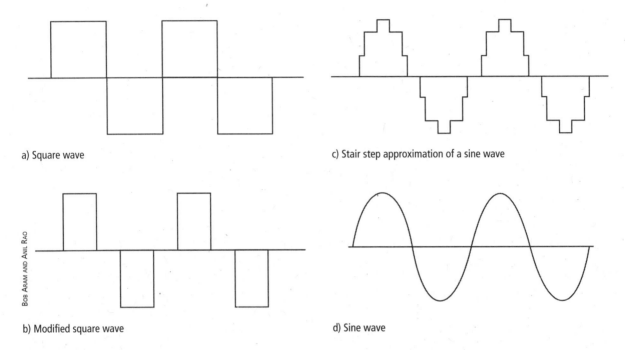

a) Square wave

c) Stair step approximation of a sine wave

b) Modified square wave

d) Sine wave

BOB ARAM AND ANIL RAO

Fig. 6-2 a, b, c and d: *Square Wave, Modified Square Wave, Stair Step and Sine Wave Electricity.*

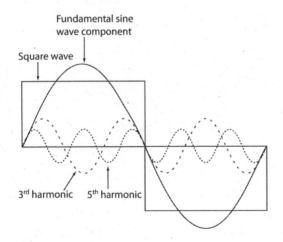

Fundamental sine wave component

Square wave

3rd harmonic   5th harmonic

Fig. 6-3: *Total Harmonic Distortion. In a square wave, and to a lesser extent modified square wave, electricity contain harmonics, waveforms at frequencies that are multiples of the desired frequency, shown here. Motors, transformers and many electronic devices do not perform well with such input.*

of electricity. If this square wave is modified so that it pauses at zero volts briefly before reversing direction, the result is *modified square wave electricity,* as shown in Figure 6-2b. Although this is closer to a *sine wave,* the kind used in most homes, it is still not suitable for use in homes and businesses.

The problem with square wave, and to a lesser extent modified square wave, electricity is that they contain *harmonics.* Harmonics are waveforms at frequencies that are multiples of the desired frequency. Square wave electricity contains odd harmonics — that is, frequencies that are 3-times, 5-times, 7-times, etc. the fundamental frequency of the square wave (Figure 6-3). The output of a 60 Hz square wave or modified square wave inverter, for

How Transformers Work

Transformers are used to increase (step up) and decrease (step down) voltage in the electrical grid and electrical devices. If you're tied to the electrical grid, chances are there's a transformer on a pole or pad outside your home (Figure 6-4). This is a step-down transformer that decreases the voltage of the electricity in the electric line running by your home, which ranges (depending on the location) from 7,500 to 22,500 volts, to 120 or 240 volts, used in homes and businesses.

Step-up transformers, like the one found in an inverter, do just the opposite. That is, they increase the voltage. Both step-up and step-down transformers use *electromagnetic induction*.

Electromagnetic induction is the production of an electric current in a wire as the wire moves through a magnetic field. This amazing phenomenon is the basis of virtually all electrical production technologies from coal- and natural-gas-fired power plants to wind turbines. For example, in the wind turbine shown in Figure 6-5, magnets are mounted on a rotating can, known as a *rotor*. It spins when the blades of the turbine spin. The magnets of a wind turbine rotate around a set of windings, coils of copper wire. They are the stationary portion of a wind turbine generator, known as the *stator*. The rotation of magnets past the stationary (stator) windings produces electric current in the windings via electromagnetic induction. Cutting through the magnetic field literally pushes electrons in the copper wire, creating a current.

Magnetic fields are also produced as electricity flows through conductors. In conductors carrying alternating current, the magnetic field expands, collapses, and then reverses direction as the flow of electrons alternates. This alternating magnetic field can induce an alternating current in a nearby wire, or winding.

Transformers consist of two wire coils, called *windings*, one that's fed electricity, and another that produces electricity at a higher or lower voltage. AC electricity flowing through the first coil creates an oscillating magnetic field. It induces an electrical current in the second coil. But how does a transformer increase or decrease voltage?

The voltage in the second coil increases and decreases in proportion to the number of turns ☞

Fig. 6-4: *Step Down Transformer. This device reduces voltage of the electricity entering your home from standard line voltage (which ranges from 7,500 to 22,500 volts) to household voltage, 120 and 240 volts.*

DAN CHIRAS

in the second winding compared to the number in the first.[2] The greater the number of turns in the second winding, the higher the voltage. If the number of turns in the second coil (the one in which an electric current is induced) is greater than the first coil, voltage in the second coil increases. If the number of turns in the second coil is lower, the voltage in the second coil is lower. ■

Fig. 6-5: *Anatomy of a Wind Turbine.*

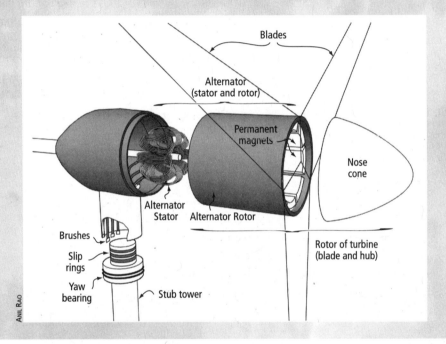

example, contains not only 60 cycle per second or 60 Hz, but also 180 Hz, 300 Hz, 420 Hz, etc. power. Motors, transformers and many electronic devices do not perform well under such circumstances. They can even be damaged by this phenomenon called harmonic distortion (see sidebar, "What is Total Harmonic Distortion?").

Filtering the output of a square wave or modified square wave inverter will reduce these unwanted harmonics, but not eliminate them.

Modified square wave electricity works in most electrical devices; however, most of them run less efficiently on this type of electricity. Because of this, most inverters sold in North America and Europe are designed to produce a much purer form of electricity known in the trade as *sine wave electricity*. It is nearly identical to the type of electricity delivered via the

What is Total Harmonic Distortion?

Total harmonic distortion (THD) is a measure of the harmonics in an inverter output — the frequencies we don't want — compared to the frequency we do want. Electricity from a square wave inverter has a THD of about 48%. The output of a modified square wave inverter is somewhat lower than this, although you won't find a value on the inverter specification sheet. (Inverter manufacturers don't like to talk about the THD of their modified square wave inverters.)

In sharp contrast, the THD of the output of a sine wave inverter is 5% or less, about the same as the distortion in utility power. In some cases, the THD of a sine wave inverter is lower than in utility power. The amount of distortion on power lines is dependent on where you live and what your neighbors — and the utility — are doing. Harmonic distortion is produced by variable speed drives for electric motors in factories and electric welders, arc furnaces, fluorescent lights and even computers. They all produce short bursts of noise or distortion in the line. In homes and small businesses, harmonic distortion is produced by fluorescent lights, computers and some power cubes. Harmonic distortion also arises from electric utility systems from overloaded transformers and other technologies. A perfect sine wave has 0% THD.

utility grid. How do sine wave inverters produce such clean power?

Sine wave inverters contain multiple H-bridges, each of which produces square wave electricity (Figure 6-6). However, the H-bridges are stacked on top of one another, that is, carefully spaced to produce a stair step waveform that closely resembles a sine wave, as shown in Figure 6-3c. Filtering this waveform produces a very clean sine wave, the kind of electricity our modern homes require (Figure 6-3d). As noted earlier, a pure sine wave has no harmonics. All the energy is in the frequency we want.

Types of Inverters

Inverters come in three basic types: those suitable for grid-connected systems, those designed solely for off-grid systems, and those made for grid-connected systems with battery backup. Table 6-1 lists alternative names for inverters and PV systems that you may encounter.

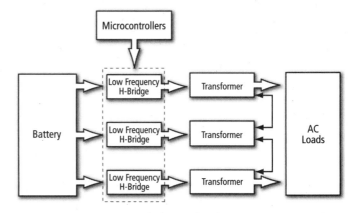

Fig. 6-6: *Multiple H-bridges are found in sine wave inverters. Each produces square wave electricity. Because the H-bridges are spaced to produce a stair step waveform that closely resembles a sine wave, filtering this waveform produces a very clean sine wave.*

Modified Square Wave or Modified Sine Wave?

A modified square wave inverter modifies — or cleans up — the square wave output from the H-bridge to reduce harmonics. Inverter manufacturers, however, usually label their modified square wave inverters as modified sine wave inverters. Why? It's a marketing ploy. Customers know that a sine wave is better than a square wave, so an inverter that produces modified sine wave sounds better than one described as producing a modified square wave. In reality, they are the same.

Grid-Connected Inverters

Today, the vast majority of renewable energy systems — both solar electric and wind — are grid-connected. These systems require inverters that operate in sync with the utility grid. They produce grid-compatible sine wave AC electricity and back feed the surplus onto the grid. The electricity they produce is indistinguishable from grid power; often it's even a bit cleaner.

Grid-connected inverters, also known as *utility-tie inverters*, convert DC electricity from the PV array into AC electricity (Figure 6-7). Electricity then flows from the inverter to the

Table 6.1
PV System and Inverter Terminology

Common terminology used in this book (referring to systems and inverters)	UL 1741 (Refers to inverter)	NEC 690-2 (Referring to system)
Grid-connected	**Utility-interactive** Operates in parallel with the utility grid.	**Interactive** A solar photovoltaic system that operates in parallel with and may deliver power to an electrical production and distribution network.
Grid-connected with battery backup	**Multimode** Operates as both or either stand-alone or utility interactive.	(Interactive, not separately defined)
Off-grid	**Stand-alone** Operates independent of the utility grid.	**Stand-alone** A solar photovoltaic system that supplies power independently of an electrical production and distribution network.
		Hybrid A system composed of multiple power sources. These power sources may include photovoltaic, wind, micro-hydro generators, engine-driven generators, and others, but do not include electrical production and distribution network systems. Energy storage systems, such as batteries, do not constitute a power source for the purpose of this definition.

BOB ARAM

breaker box or main service panel. It is then fed into active circuits, powering refrigerators, computers, stereo equipment, and the like. Surplus electricity is then fed onto the grid, running the electrical meter backward.

Grid-tied inverters produce sine wave electricity that matches the grid both in frequency and voltage. To perform this function, these inverters monitor the voltage and frequency of the electricity on the utility lines. They then adjust their output to conform to the grid. That way, electricity back fed from a PV system onto utility lines is identical to the electricity utilities are transmitting to their customers.

Grid-compatible inverters are equipped with *anti-islanding protection* — a feature that automatically disconnects the inverter from the grid in case of loss of grid power. That is to say, grid-connected inverters are programmed to shut down if the grid goes down. If the utility grid goes down, the inverter stays off until service is restored. Grid-tied inverters terminate the flow of electricity to the grid for a good reason: to protect utility workers from electrical shock.

These inverters also shut down if there's an increase or decrease in either the frequency or voltage of grid power outside the inverter's acceptable limits (which are established by the utility companies). These are referred to as *over/under voltage* or *over/under frequency*. If either the voltage or the frequency varies from the settings programmed into the inverter, it turns off. This can occur at the end of a utility line or near a manufacturing plant that has lousy power factor. (Power factor was defined in Chapter 4 and is explained in

more detail in the accompanying sidebar, "How Does Power Factor Affect Voltage") In such instances, grid power is poor enough that it doesn't match the window required by the inverter. The inverter remains off until the grid power returns within the acceptable range.

Grid-connected inverters also come with a *fault condition reset* — a sensor and a switch that turns the inverter on once the grid is back up or the inverter senses the proper voltage and/or frequency.

One thing that many who are new to solar electricity often have difficulty understanding is that when the grid goes down or experiences over/under voltage or over/under frequency, not only does the inverter stop back feeding

Fig. 6-7:
Grid-connected inverter. This sleek, quiet inverter from Fronius is suitable for grid-connected systems without batteries. Fronius' inverters are produced at a new production facility in Sattledt, Austria, which started production in early 2007. The facility receives 75% of its electricity from a large, roof-mounted PV system and 80% of its heat from a biomass heating system.

surplus onto the grid, the entire PV system shuts down. That is, the flow of electricity to the main service panel and to active circuits in the house ceases as well. Why doesn't the inverter keep working, supplying AC electricity to the home and business?

Is My Inverter Safe?

How does a buyer know if their inverter will disconnect from the utility as it should in an outage? More importantly, how can he or she convince the local utility that it will? The answer is on the inverter nameplate. If it lists compliance with UL 1741 and IEEE 1547, you (and your utility) can be assured that the inverter meets national safety standards — that is, it ensures safe interconnection of distributed generation sources. Most inverters today are UL 1741 certified, which should smooth the interconnection process with any local utility.

How Does Power Factor Affect Voltage?

Power factor is a measure of how well the voltage and the current waveforms in electrical current are aligned. Inductive loads such as motors cause the current to lag behind the voltage. When the voltage and current waveforms are out of alignment, the real power delivered decreases. To deliver a given amount of power, the current must increase as power factor increases. Current flowing in wires causes voltage drop, which lowers the voltage at the end of the line. A factory with a "lousy power factor" requires far higher current than is needed to power loads with a good power factor. This higher current may cause the voltage in the power distribution line to drop.

The answer's simple: because the inverter needs the grid connection to determine the frequency and voltage of the AC electricity it produces. Without the connection, the inverter can't operate. It's just wired that way.

To avoid losing power when the grid goes down, you can install a grid-connected system with battery backup. These systems allow homes and businesses to continue to operate, even though the utility grid has failed. Although inverters in such systems disconnect from the utility during outages, they continue to draw electricity from the battery bank to supply active circuits. They do so via a separate breaker box. Such systems are typically designed and wired to provide electricity only to essential circuits in a home or business, supplying the most important (critical) loads. Although a homeowner or business owner may see a momentary power flicker when the grid goes down and the inverter switches to battery operation, the better the inverter, the smoother the transfer. High-quality inverters switch so fast that even computers are not affected when the system shifts to battery backup.

Grid-connected inverters also frequently contain LCD displays that provide information on the input voltage (the voltage of the DC electricity from the array) and the output voltage (the voltage of the AC electricity the inverter produces and delivers to a home and the grid). They also display the current (amps) of the AC output. Fronius inverters also keep tab of the dollars saved and pounds of carbon dioxide emissions avoided by generating solar electricity.[3]

Another key feature found in many modern inverters is a function known as *maximum power point tracking* (MPPT) (also discussed in Chapter 4). MPPT is software that continually tracks a solar array's voltage and amperage. It does so to find the combination of amps and volts that result in the maximum wattage, known as the maximum power point. This ensures that the solar system produces the most power (watts) while in operation, improving the efficiency and optimizing the output. MPPT is also found in charge controllers in battery-based systems. The accompanying sidebar, "Maximum Power Point Tracking" explains how maximum power point tracking works.

Some inverters, like the Trace SW utility-intertie series inverters, come with automatic morning wake-up and evening shutdown. These features shut the inverter down at night (as it is no longer needed) and wake it up in the morning (to get ready to start converting DC electricity from the array into AC electricity.) This sleep mode in the SW series inverters uses less than 1 watt of power.

Grid-connected inverters operate with a fairly wide input range. The DC operating range of the SW series, for instance, ranges from 34 to 75 volts DC (you might see this listed as "VDC"). Inverters from Fronius and Oregon-based PV Powered Design are designed to operate within a broader DC voltage input range: from 150 to 500 volts DC. This permits the use of a wider range of modules and system configurations. Moreover, high-voltage arrays can be placed farther from the inverter than low-voltage arrays. In addition, high-voltage DC input means that

smaller and less expensive wires can be used to transmit electricity to a home or office from the array. With the cost of copper skyrocketing as a result of higher energy prices and higher demand, savings on wire size can be substantial.

Grid-connected inverters are installed indoors and outdoors. Some, like the Trace UT series, are installed in a protective outdoor enclosure. This permits installers to mount the inverter on exterior walls, which is required by many utilities. To cool the inverter, the cabinet includes openings (ventilation louvers) and internal baffles to increase the surface area for heat loss. They can also be ordered with a cooling fan for additional cooling for installations in areas that experience high outside temperatures. Inverters such as this contain AC and DC circuit breakers that permit operators to disconnect all AC and DC power sources when servicing the unit. Integrated AC and DC disconnects reduce installation costs.

Grid-connected inverters must comply with the specifications set forth in the interconnection agreement between the utility and the producer (a.k.a. the customer). Utility specifications include such things as the acceptable distortion of waveform, over/under frequency, over/under voltage, and automatic shutdown during a power outage. When considering an inverter, look for the UL 1741 label. Inverters bearing this label should meet all of the utility requirements.

Numerous companies produce utility-intertie inverters, including Xantrex, Fronius, SMA and PV Powered Design. Inverters typically come with two- to ten-year warranties.

Maximum Power Point Tracking

To get the most out of a PV array, most systems incorporate a function known as maximum power point tracking (MPPT). The circuitry that allows MPPT is located in the inverter in grid-connected systems and in the charge controller of battery-based systems.

As noted in Chapter 3, the voltage and current produced by a PV array vary with changes in cell temperature, irradiance and load. (You'll recall that a load is any device, equipment or appliance that consumes electricity, batteries included).

The relationship between current and voltage produced by a PV device (a cell, module or array) is shown in Figure 6-8. This graph is known as a *current-voltage curve* or *I-V curve* (I represents amps and V represents volts). Take a moment to study the graph.

I-V curves show the relationship between current and voltage produced by a PV array — under a specific set of conditions, that is, a specific irradiance level and cell temperature. The PV array can operate anywhere along this curve.

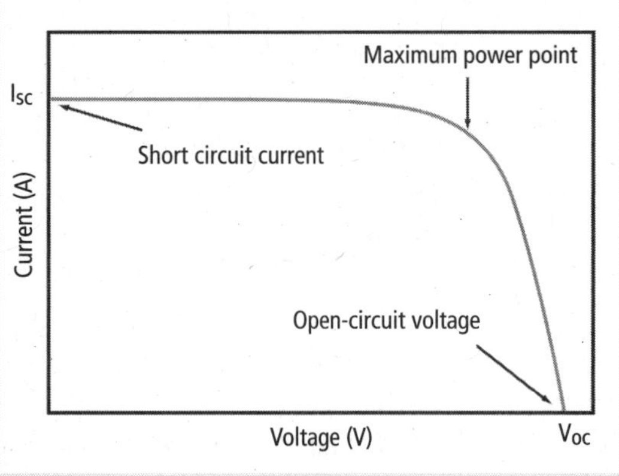

To begin, look at the point on the curve labeled "short-circuit current." This is where the array operates if its output is short-circuited, i.e., the array's terminals are connected together. As indicated on the graph, the current is high, but the voltage is zero. *No* power is being produced at this point.

Why?

As you recall, power (watts) is determined by multiplying volts by the current (amps). Since the volts are zero at this point, the power output is zero.

Now look at the point labeled "open-circuit voltage." This is the point at which the array operates if its output is open circuited, i.e., not connected to anything. The current is zero and the voltage is at a maximum. What is the power being produced at this point?

Once again, the power output (watts) is zero.

At all points in between the short-circuit and open-circuit points, power is produced. How much power an array produces depends on where the array is operating on the curve. As the operating point moves away from the short-circuit point, the power output rises. The power reaches its peak near the middle of the "knee," the area where the curve changes from flat to steep. The point on the curve where the power reaches its maximum is called the maximum power point. As the operating point moves down the "knee," away from the maximum power point and toward the open-circuit point, the power decreases. It is the *load* that determines where on its I-V curve an array will operate. ☞

Fig. 6-8: *I-V Curve*

Just as the current and voltage of the PV array can be plotted on an I-V curve, so can the current and voltage of the load. Figure 6-9 shows the I-V curve of a typical load, which is generally a straight line. The load can operate anywhere on this line — at any combination of volts and amps. So, what determines where on its I-V curve the load will operate?

The answer is the array.

When the array and the load are connected, they must operate at the same point — the one point that is common to both lines. This is illustrated by superimposing the I-V curves of the array and the load, shown in figure 6-10. Where the two lines cross determines the operating point of the array when it is connected to the load.

In order to get the maximum power out of the array, it must operate at its maximum power point. But, as we have just seen, the array will operate at a point determined by the load. So, how do we get the load line to intersect the array curve at the maximum power point?

Before we answer this question, we need to point out that the array I-V curve is not static. It changes with cell temperature and irradiance. Figure 6-11 shows how the array I-V curve changes with cell temperature. Figure 6-12 shows how the array I-V curve changes with irradiance. Because of this, the array maximum power point is a moving target. To get the load line to intersect the array curve at the maximum power point, we need to alter the load line to "track" the moving maximum power point of the array I-V curve. Fortunately, there is a way to do this.

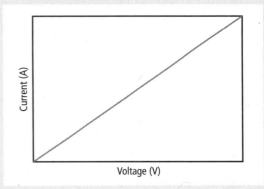

Fig. 6-9: *I-V Curve (load).*

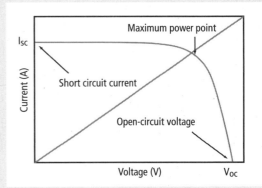

Fig. 6-10: *I-V Curve for PV Array and Load*

The slope, or angle, of the load line can be altered by changing the *impedance* of the load. Impedance is the resistance to the flow of current in a conductor and is measured in *ohms*. The impedance of the load is calculated by dividing the voltage by the current. Impedance changes via a circuit known as a *DC-to-DC converter,* which ☞

can change the voltage of the electricity traveling from the array to the load (the charge controller or inverter). As the voltage changes, the current also changes. As a result, the DC-to-DC converter can change the slope of the load line. Additional circuitry can control the DC-to-DC converter to "aim" the load line so that it intersects the array curve at the maximum power point. The combination of the DC-to-DC converter and the circuitry that controls it is known as a maximum power point tracker.

As temperature, irradiance and load change during the day, the MPPT adjusts the impedance of the load, ensuring maximum power production by a PV array. The MPPT can also reduce the voltage of a high-voltage array to match the lower voltage needed by a battery or grid-connected inverter. ■

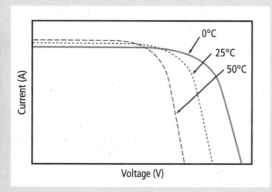

Fig. 6-11: *I-V Curve at Various Temperatures.*

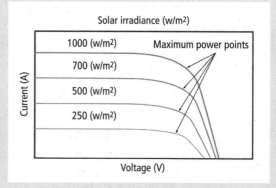

Fig. 6-12: *I-V Curve vs. Irradiance*

Fronius inverters come with a remarkable 10-year warranty.

Off-Grid Inverters

Battery-based inverters were once the mainstay of the renewable energy business, as most PV systems were for remote off-grid applications. Over time, though, as solar electricity entered the mainstream, grid-connected systems have come to dominate the market. Nevertheless, there are a few companies that produce excellent off-grid inverters (Figure 6-13).

Like grid-connected inverters, off-grid inverters convert DC electricity into AC and boost the voltage to 120 or 240 volts. Off-grid inverters also perform a number of other functions as well, discussed in this section. If you're installing an off-grid system, be sure to read this carefully.

Battery Charging

Battery-based inverters used in off-grid and grid-connected systems with battery backup typically contain battery chargers. (Batteryless grid-tied inverters don't.) As their name implies, battery chargers charge batteries from an external source — either a gen-set in an off-grid system or the grid in a grid-connected system with battery backup. But isn't the battery charged by the PV array through the charge controller?

The charge controller in battery-based systems is often referred to as a programmable battery charger, because it does indeed charge batteries. However, it is more accurate to refer to the device as a battery charge *regulator*. Why?

The charge controller receives DC electricity from a solar electric array and then sends it to the battery bank. The main function of the charge controller, however, is to regulate the charging of the battery bank. That is, the charge controller regulates the flow of the electricity from the array to the battery bank. (More on this in Chapter 7.) A battery charger in the inverter, on the other hand, takes AC from a gen-set or the grid and converts it to DC. It then feeds DC electricity to the batteries.

In off-grid systems, battery charging gensets are used to restore battery charge after periods of deep discharge. This prolongs battery life and prevents irreparable damage to the plates. Battery chargers are also used during equalization. You'll learn about both of these processes and batteries in Chapter 7.

Abnormal Voltage Protection

High-quality battery-based inverters also contain programmable high- and low-voltage disconnect features. These features protect various components of a system, such as the

XANTREX (XW SERIES INVERTER)

Fig. 6-13: *Battery-Based Inverter. This inverter from Xantrex can be used for grid-connected systems with batteries and off-grid systems.*

batteries, appliances and electronics in a home or business. They also protect the inverters.

The low-voltage disconnect (LVD) in an inverter monitors battery voltage at all times. When low battery voltage is detected (indicating the batteries are deeply discharged) the inverter shuts off and often sounds an alarm. The flow of electricity from the batteries to the inverter stops. The inverter stays off until the batteries are recharged.

Low-voltage disconnect features are designed to protect batteries from very deep discharging, which can damage batteries. Although lead-acid batteries are designed to withstand deep discharges, discharging batteries beyond the 80% mark causes irreparable damage to the lead plates inside a battery and leads to a battery's early demise. Although complete system shutdown can be a nuisance, the disconnect feature is vital to the long-term health of a battery bank.

Although the low-voltage disconnect feature is critical, *Home Power*'s Ian Woofenden notes that "it should not be used to manage one's batteries. It is designed more to protect an inverter from low voltage than to protect batteries from deep discharge. If the voltage of the battery bank reaches the disconnect level," he adds, "you may have already damaged your batteries." It's much better to develop an awareness of the state of charge of your batteries and learn to modify usage or turn on a generator well before the low voltage disconnect kicks in.

To avoid the hassle of having to manually start a generator when batteries are deeply discharged, some inverters contain a sensor and switch that activates the generator

A Cautionary Note

Home Power's Ian Woofenden notes that some renewable energy experts are leery about using auto-start generators. They prefer to have a human being in charge of the system, monitoring the charge of the batteries and the condition of the generator, including its fuel supply. Why not automate? One reason is that if the inverter malfunctions, and it fails to shut down the gen-set when the batteries are fully charged, batteries can be overcharged and ruined. Generators not equipped with low-oil protection (a feature that turns the generator off if the oil levels are too low) can also be ruined if they're run when oil levels are low. Also, fuel levels should be monitored because the generator should not run out of gas while charging a battery — and automated systems don't have low-fuel protection.

automatically. When low battery voltage is detected, the inverter sends a signal to start the generator, provided the generator has a remote start capability. The fossil fuel generator then recharges the batteries. AC electricity from the gen-set is converted to DC electricity by the inverter's battery charger. It then flows to the batteries, charging them.

When the generator is operating, the inverter also sends some of the AC electricity it produces to the main service panel (in off-grid systems) or to a critical loads panel (in grid-connected systems) to supply active loads. In grid-connected systems with battery backup, the grid serves the same function as the backup generator, although some grid-connected systems with battery backup also include gen-sets to recharge batteries during prolonged power outages. As you might suspect, more sophisticated auto-start generators cost more than the standard pull-cord type.

High-Voltage Protection

Inverters in battery-based systems also often contain a high-voltage shut-off feature. This sensor/switch terminates the flow of electricity from the gen-set when the battery voltage is extremely high. (Remember: high battery voltages indicate that the batteries are full.) High-voltage protection therefore prevents overcharging, which can severely damage the lead plates in batteries. It also protects the inverter from excessive battery voltage.

Inverters designed for off-grid systems cost a little more than the less complicated grid-tied inverters. However, the most significant savings in a batteryless grid-connected system comes from the omission of batteries and a battery room. (A very small battery bank can easily cost $1,500 to $2,500; most battery banks cost twice that.) Batteries need replacement every five to ten years, depending on the type of battery you install and how well they are treated. Deep cycling and allowing the battery to sit partially discharged for long periods will shorten their useful life. Avoiding batteries, therefore, reduces costly battery replacement. It also saves on the time required to maintain batteries, including equalization — a rejuvenating procedure we'll discuss in detail in the next chapter. When comparing costs of an off-grid system to a grid-connected system, don't forget to factor in line extension and utility connection costs for the latter. Battery backup may be a much cheaper option in the long run, if it avoids a $20,000 to $50,000 grid-connection fee.

Multifunction Inverters

Grid-connected systems with battery backup are popular for individuals with homes that can't afford to be without electricity for a moment. They can also be crucial for businesses. These applications require multifunction inverters — battery- and grid-compatible sine wave inverters. They're commonly referred to as *multifunction* or, less commonly, *multimode inverters*. Both applications require a charge controller.

Multifunction inverters contain features of grid-connected and off-grid inverters. Like a grid-connected inverter, they contain anti-islanding protection that automatically disconnects the inverter from the grid in case

of loss of grid power, over/under voltage, and over/under frequency. They also contain fault condition reset — to power up an inverter when a problem with the utility grid is corrected. Like grid-connected inverters, multifunction inverters contain battery chargers and high- and low-voltage disconnects.

Grid-connected systems with battery backup are the most difficult to understand. Many people think that the batteries are used to supply electricity at night or during periods of excess demand — that is, when household loads exceed demand. That's not true. The grid is the source of electricity at night or during periods when demand exceeds the output of the PV array. The batteries are there only in case the grid goes down.

How electricity flows during the day depends on the state of charge of the batteries. If the batteries are full, DC electricity from the array travels from the array through the charge controller and then to the inverter. The inverter in such instances acts like a load diversion charge controller (described in Chapter 7) by taking all of the energy from the array, beyond what it takes to hold the battery at a float/full charge, and powers all of the critical and non-critical loads in the house. It sends excess through the building's meter and out onto the electrical grid. It is important to note that the charge regulating set points on the charge controller have to be set higher than the set points of the inverter, so that the charge controller only comes into play in the event of a utility power outage, when the inverter can no longer hold battery voltage to the float voltage set point by selling to the grid.

If the batteries run low, for example, after a utility outage, the inverter charges them and also powers active loads — both from the PV array (if the sun is shining) and from electricity drawn from the utility lines. Once the battery bank is full, the system returns to normal operation. At night, the home is supplied by AC electricity from the utility grid, not the batteries, unless there is a power outage.

If you are installing an off-grid system, you may want to consider installing a multifunction inverter in case you decide to connect to the grid in the future. For those who want to install a grid-connected system with battery backup remember that although multifunction inverters allow system flexibility, they are not the most efficient inverters. That's because some portion of the electricity generated in such a system must be used to keep the batteries topped off. This may only require a few percent, but over time a few percent add up. In systems with poorly designed inverters or large backup battery banks, the electricity required to maintain the batteries can be substantial. It is also worth noting that as batteries age, they become less efficient; more power is consumed just to maintain the float charge, which reduces the efficiency of the system.

If feeding electricity back onto the grid is one of your objectives, be careful when choosing an inverter. For example, Xantrex SW-series inverters are not optimized for grid sell-back. They focus on keeping the batteries full. Backfeed is a second priority. For best results, we recommend inverters that prioritize the delivery of surplus to the grid while preventing deep discharge of the battery bank. "OutBack's inverters are a big step forward in this respect," notes Ian Woofenden, "and other manufacturers (including Xantrex's newer models — their XW series) are improving their products as well."

If you want the security of battery backup in a grid-connected system, we suggest that you isolate and power only critical loads from the battery bank. This minimizes the size of your battery bank, reduces system losses, and reduces costs. Unless you suffer frequent or sustained utility outages, a batteryless grid-connected system usually makes more sense from both an economic and environmental perspective.

With these basics in mind, we turn our attention to the purchase of an inverter.

Buying an Inverter

Inverters come in many shapes, sizes and prices. The smallest inverters, referred to as *pocket inverters,* range from 50 to 200 watts. They are ideal for supplying small loads such as VCRs, computers, radios, televisions and the like. Most homes and small businesses, however, require inverters in the 2,500 to 5,500-watt range. Which inverter should you select?

Type of Inverter

Professional solar installers will design a system for you, and specify the inverter and other components. For those interested in designing and installing their own systems (a project best undertaken by those with electrical experience and a few PV installation workshops under their belts), the first consideration when shopping for an inverter is the type of system you are installing. As noted earlier, if you are installing an off-grid system, you'll need a battery-based inverter and a charge controller. If you are installing a grid-connected system with no batteries, you'll need an inverter designed for grid connection. The inverter must be matched to the output of the solar array. (Installers can help match the inverter to the array.) If you are installing a grid-connected system with battery backup, you must purchase an inverter that is both grid- and battery-compatible. Battery-based grid-tied inverters may also be a good choice for off-grid systems, in case you decide to connect to the electrical grid at a later date. Two such inverters are the OutBack GFX or the Xantrex XW series inverters (they replaced the SW series starting in 2007).

When purchasing an inverter through an installer, your choices may be limited because most installers carry a line of inverters they know and like. An installer will very likely be a dealer for a line of inverters. Installers therefore make a recommendation that fits your needs from the inverters they carry, saving you the research required to select the right inverter.

Even though a local supplier/installer may recommend an inverter, reading the following material will expand your understanding of inverters. It will also help you better manage and troubleshoot your renewable energy system. If you are designing and installing your own system, you'll surely want to know as much as you can about inverters. Here are some important things to look for in an inverter.

System Voltage

When shopping for a battery-based inverter, you'll need to select one whose input voltage corresponds with the battery voltage of your system. As noted earlier, wind, solar electric and micro hydro systems are typically either 12-, 24- or 48-volt systems. Twelve-volt systems are common in small applications, such as cabins or summer cottages. Residential systems are either 24- or 48-volt systems, most commonly 48 volts, because they're the most efficient.

System voltage refers to the voltage of the electricity produced by a renewable energy technology — a solar electric array, a wind turbine or micro hydro generator. That is, the PV arrays or the generators in wind turbines are wired to produce 12-, 24- or 48-volt electricity. The batteries are wired similarly, so that they produce 12-, 24- or 48-volt electricity. Inverters, in turn, increase the voltage to 120 or 240 volts AC. Because all components of an off-grid renewable energy system must operate at the same voltage, the inverter must match the source and battery voltage. A 24-volt inverter won't work in a 48-volt system. If you are installing a 48-volt array, you'll need a 48-volt battery-based inverter. You must also wire your battery bank for 48-volts. If installing a 24-volt array, you'll need a 24-volt battery-based inverter and battery bank. It is always a good idea to talk with the manufacturer's technical staff to ask for their input on the best inverter.

When shopping for an inverter for a grid-connected system, you'll need to find one that matches the output of your proposed array, as explained in the accompanying sidebar, "PV Array Voltage, String Size, and Choosing the Correct Inverter."

Modified Square Wave vs. Sine Wave

The next inverter selection criterion you must consider is the output waveform. As noted earlier in this chapter, battery-based inverters

PV Array Voltage, String Size, and Choosing the Correct Inverter

In the 1970s through the mid 1990s, most PV systems were designed to charge 12-, 24- or 48-volt battery banks in off-grid systems, the predominant type of PV system being installed. To meet the demands of this market, manufacturers produced modules rated at 12 volts and installers built systems around them. In a 12-volt system, for instance, an installer would wire one or more 12-volt modules in parallel to produce the voltage needed to charge a 12-volt battery bank. In 48-volt systems, installers would wire four 12-volt modules in series, then install one or more strings (a string is a group of modules wired in series) in parallel to produce 48-volt DC electricity. (Wiring modules in series increases the voltage; wiring in parallel increases the amperage.)

The 12-, 24- or 48-volt DC electricity was then fed to 12-, 24- or 48-volt charge controllers. They deliver the electricity to appropriately sized battery banks. Appropriate inverters were installed to draw off the 12-, 24- or 48-volt battery banks. They converted the low-voltage DC electricity into 120-volt AC electricity suitable for household use.

Today, systems based on 12-volt increments have largely fallen by the wayside. One reason for this is that most modern PV systems are batteryless grid-connected. Because there are no batteries, the inverters in these systems can be designed to accept DC electricity from arrays at much higher voltages. In fact, most of grid-connected inverters operate with an input range of 150 to 550 volts with a maximum of 600 volts. (The National Electrical Code prohibits wiring PV arrays for homes higher than 600 volts.)

Higher voltage reduces losses as electricity flows from the array to the inverter. Because less current flows at higher voltage, higher voltage systems permit the use of smaller gauge conductors, which reduces installation costs. The higher voltage also allows smaller components in the inverters, reducing their size, weight and cost.

To accommodate the higher voltage inverters, manufacturers now produce PV modules with higher nominal voltages — 16, 24 and 36 volts. They are wired in series strings to produce high-voltage DC electricity for today's high-voltage inverters.

To help installers determine how many modules they can wire in a string, virtually all inverter manufacturers provide online calculators. To calculate string size, the installer simply enters the temperature conditions of the site and the type of module he or she is considering. The online calculator provides the maximum, minimum, and ideal number of modules in a series string for each of its inverters.

Temperature is important when sizing an array because output changes with temperature. ☞

are available in two types: modified square wave (often called modified sine wave) and sine wave. Grid-connected inverters are all sine wave so they match utility power.

Modified square wave electricity is a crude approximation of grid power but works well in many appliances such as refrigerators and washing machines and in power tools. It also works

---

Low temperatures, for example, increase the output of an array. If an array has not been sized carefully, the voltage of the incoming electricity could exceed the rated capacity of the inverter and the charge controllers — if it is a battery-based system. High voltage can damage this equipment. (This is most likely to occur early on cold, sunny days when the voltage is highest.)

High temperatures, in contrast, reduce the output of an array. If the voltage of a PV array falls below the rated input of a direct grid-tied inverter, a PV system will shut down and will remain off until the array cools down and the voltage increases. This, of course, reduces the efficiency of a system.

Many modern battery-based systems can be designed to handle higher voltage arrays thanks to the introduction of maximum power point tracking (MPPT) charge controllers. MPPT charge controllers optimize power output of an array and accept higher voltage input from PV arrays, but they contain step-down transformers that reduce the high-voltage array input to charge the battery bank at the appropriate voltage. These systems are generally designed with an open-circuit voltage of around 150 volts, and can be used to charge 12-, 24- or 48-volt battery banks. (Open-circuit voltage is the voltage that is generated by a module with no amps flowing, usually first thing in the morning,

during which time there's just not enough sunlight to produce useful current flow through the charge controller.)

A PV array's operating voltage is influenced by the ambient temperature, as just noted, and also the number of modules. Array temperature is also influenced by the type of mounting. As you'll learn in Chapter 8, some array mounts allow more air to circulate around modules than others. Other mounting options place PV modules in arrays very close to hot roof surfaces, which experience higher temperatures. The hotter the array, the greater the decline in its output. When calculating string size, then, an installer must also stipulate the type of mounting.

Designers must also take into account the slow degradation of modules, which reduces their output over time. (Module output may decline as a result of moisture that seeps into the PV cells, which corrodes the internal electrical connections. Module deterioration also results from the slow deterioration of the ethylene vinyl acetate coating on the PV cells, discussed in Chapter 3.) Because of this, it is important *not* to design a system that operates consistently near the lower end of the inverter's input voltage window. As the modules' output declines over time, you'll increase the chances of reduced energy output on hot days. ■

pretty well in most electrical devices, including TVs, lights, stereos, computers and inkjet printers. Although all these devices can operate on this lower quality waveform, they all run less efficiently, producing more heat and less work — light or water pumped, etc. — for a given input.

Problems arise, however, when modified square wave is fed into sensitive electronic circuitry such as microprocessor-controlled, front-loading washing machines, appliances with digital clocks, chargers for various cordless tool, copiers and laser printers. These devices require sine wave electricity to operate. Without it, you're sunk. Dan, for example, found that his energy-efficient front-loading Frigidaire Gallery washing machine would not run on his modified square wave electricity produced by his very first inverter. The microprocessor that controls this washing machine — and other similar models (except the Staber) — simply can't operate on this inferior form of electricity. Since he replaced this inverter with a sine wave inverter, Dan's had no troubles whatsoever.

Certain laser printers may also perform poorly with modified square wave electricity. The same goes for some battery tool chargers, ceiling fans and dimmer switches.

Making matters worse, some electronic equipment such as TVs and stereos give off an annoying high-pitched buzz or hum when operating on modified square wave electricity. Modified square wave electricity may also produce annoying lines on TV sets, and can damage sensitive electronic equipment.

When operated on modified square wave electricity, microwave ovens cook slower.

Equipment and appliances also run warmer and might last fewer years on modified square wave electricity. Computers and other digital devices operate with more errors and crashes. Digital clocks don't maintain their settings as well. Motors don't always operate at their intended speeds. So why do manufacturers produce modified square wave inverters?

The most important reason is cost. Modified square wave inverters are much cheaper than sine wave inverters. You will most likely pay 30 to 50% less for one.

Another reason for the continued production of modified square wave is that they are hardy beasts. They work hard for many years with very little, if any, maintenance. (Their durability may be related to their simplicity: they are much less complex electronically than sine wave inverters.)

Modified square wave inverters come in two varieties: high frequency switching and low-frequency switching units. High-frequency switching units are the cheaper of the two. A typical 2,000-watt high-frequency switching inverter, for example, costs 20 to 50% less than low frequency models. They are also much lighter than low-frequency switching models, and are, therefore, easier to install. The high-frequency inverter may weigh 13 pounds compared to the 50 pounds of a low-frequency inverter.

Low-frequency switching modified square wave inverters cost more and weigh more than their high-frequency switching cousins, but are well worth the investment. One reason for this is that they typically have a much higher surge capacity. That means they can deliver greater surges of power needed to start certain electrical devices such as well pumps, power tools, dishwashers, washing machines and refrigerators. More on surge capacity shortly.

Although there are some reliable modified square wave inverters on the market, we recommend that you purchase a sine wave battery-based inverter for off-grid systems. Their output is well suited for use in modern homes with their array of sensitive electronic equipment. SMA, Xantrex and OutBack all produce excellent sine wave inverters at a reasonable price.

Output Power, Surge Capacity and Efficiency

When selecting an inverter, even a grid-tied inverter, you will also need to consider three additional factors: continuous output, surge capacity and efficiency.

Continuous Output

Continuous output is a measure of the power an inverter can produce on a continuous basis — provided there's enough energy available in the system. The power output of an inverter is measured in watts, although some inverter spec sheets also list continuous output in amps (to convert watts to amps use the formula watts = amps x volts). Xantrex's sine wave inverter, model SW2524, for instance, produces 2,500 watts of continuous power. This inverter can power a microwave using 1,000 watts, an electric hair dryer using 1,200 watts, and several smaller loads simultaneously without a hitch. (By the way, the "25" in the

Reliable High-Frequency Inverters

Although there are advantages to conventional low-frequency switching (transformer-based) inverters, there are some reliable high-frequency switching inverters on the market. Statpower and Exeltech, for example, manufacture reliable lightweight high-frequency switching inverters. You'll also find some reasonably good quality Asian inverters on the market that are well supported in North America, such as Samlex.

model number indicates the unit's continuous power output; it stands for 2,500 watts. The "24" indicates that this model is designed for a 24-volt PV system.) The spec sheet on this inverter lists the continuous output as 21 amps.

OutBack's sine wave inverter VFX3524 produces 3,500 watts of continuous power and is designed for use in 24-volt systems. Most homes can easily get by on a 3,000- to 4,000-watt inverter. Homes rarely require the full output.

To determine how much continuous output you'll need, add up the wattages of the common appliances you think will be operating at once. Be reasonable, though. Typically, only two or three large loads operate simultaneously. However, remember that, in some instances, multiple loads can operate simultaneously. A washer and well pump may be operating, for example, when a dryer and sump pump are running. If you are planning to operate a shop next to your home on the same inverter, you may need an inverter with a higher continuous output.

Surge Capacity

Electrical devices with motors, such as vacuum cleaners, refrigerators, washing machines and power tools, require a surge of power to start up. To observe this, simply watch the amp meter in a renewable energy system when someone turns on a device such as a vacuum cleaner. Or watch the amperage on a Watts Up? or Kill A Watt meter, described in Chapter 4. If you pay close attention, you'll see a momentary spike in amperage. That's the power surge required to get the motor running. The spike typically lasts only a fraction of a second. Even so, if an inverter doesn't provide sufficient power, the power tool or appliance won't start! Not starting is not just inconvenient. Stalled motors draw excessive current and overheat very quickly. Unless they are protected with a thermal cutout, they may burn out.

When shopping for an inverter, be sure to check out the surge capacity. All quality low-frequency inverters are designed to permit a large surge of power over a short period, usually about five seconds. Surge capacity or surge power is listed on spec sheets in either watts or amps.

Efficiency

Another factor to consider when shopping for inverters is their efficiency. Efficiency is calculated by dividing the energy coming out of an inverter by the energy going in.

While inverter efficiencies range from 80 to 95%, most models boast efficiencies in the low 90s. A 92% efficient inverter, for instance, loses 8% of the energy it converts from DC to 120-volt AC.

It should be noted that the efficiency of inverters varies with load. Generally, an inverter achieves its highest efficiency once output reaches 20 to 30% of its rated capacity, according to Richard Perez, renewable energy expert and publisher of *Home Power* magazine. A 4,000-watt inverter, for instance, will be most efficient at outputs above 800 to 1,200 watts. At lower outputs, efficiency is dramatically reduced.

Cooling an Inverter

Because they're not 100% efficient, inverters produce waste heat internally. A 4,000-watt inverter running at rated output at 90% efficiency produces 400 watts of heat internally. (That's equivalent to the heat produced by four 100-watt light bulbs.) The inverter must get rid of this heat or it will be damaged.

Inverters rely on cooling fins and fans to rid themselves of excess heat to reduce internal temperature. Fins increase the surface area from which heat can escape. Fans blow air over internal components, stripping off heat. If the inverter is in a hot environment, however, it may be difficult for it to dissipate heat quickly enough to maintain a safe temperature.

Inverters are also often equipped with circuitry that protects them from excessive temperature. That is, they're programmed to power down (produce less electricity) when their internal temperature rises above a certain point. As the internal temperature of the power oscillator (H-bridge) increases, the output current decreases. Translated, that means if you need a lot of power, and the internal temperature of your inverter is high, you won't get it. A Xantrex SW series inverter, for example, produces 100% of its continuous power at 77°F (25°C), but drops to 60% at 117.5°F (47.5°C).

The implications of this are many. First, inverters should be installed in relatively cool locations. Dan's is mounted in the coolest part of his earth-sheltered house, his utility room. If the inverter is mounted outside, make sure it is shaded. Second, inverters should be installed so that air can move freely around them. Don't box an inverter in to block noise (though not all inverters are noisy). Third, be sure to purchase an inverter that does a good job of dissipating heat. Ideally, you want an inverter that cools itself passively — one that doesn't require a cooling fan that consumes electricity. Many battery-based inverters have fans that cool them, for example, during battery equalization.

OutBack's inverters have a die-cast aluminum casing with finned sides that dissipate internal heat passively (Figure 6-14). In some models, the internal circuitry is completely sealed off from the outside, protecting internal components from dust, moisture, insects and small rodents that can doom an inverter. Their vented models, however, allow outside air to flow through the internal electronics, improving performance. These units produce more AC power than the sealed inverters and thus perform better in hot environments.

Battery Charger

Battery chargers are standard in most battery-based inverters designed for homes. So, if you are installing an off-grid system or a

grid-connected system with batteries, be sure to look for one with a battery charger. This will allow you to charge your DC battery bank using AC from the utility (for grid-connected systems with battery backup) or a generator (for either off-grid systems or grid-connected systems with battery backup).

Whatever inverter you get, bear in mind that the battery charger in the inverter must be capable of charging the battery in a reasonable time. As well, discussed in Chapter 7, the generator must be sized to provide the current the charger needs *plus* extra current to run household loads during charging.

Inverters can be installed inside a home or outside. While many inverters are contained in weather-tight enclosures, all inverters need to be protected from very low and very high temperatures, as mentioned previously. Many

Fig. 6-14: *The fins on this Outback inverter dissipate heat generated internally.*

direct grid-tied inverters are rated for installations with ambient air temperatures above -10° F. Battery-based inverters are typically installed inside, close to the batteries to reduce line loss. (As noted, the batteries also need to "live" in a warm space.) Grid-tied inverters are almost always installed inside the house near the main circuit breaker panel. That's where the utility service enters a house. (Most inverter manufacturers recommend their equipment be housed at room temperature.)

Noise and Other Considerations

If you are planning on installing an inverter inside your business or home, be sure to check out the noise it produces. Inquire about this upfront, and, better yet, ask to listen to the model you are considering in operation to be sure it's quiet. Don't take a manufacturer's word for it. Dan's Trace inverter (now manufactured by Xantrex) is described by the manufacturer as "quiet," but it emits an annoyingly loud buzz. The first six months after he moved into his home, the inverter's buzz drove him nuts, but he has grown used to it.

Some folks are also concerned about the potential health effects of extremely low frequency electromagnetic waves emitted by inverters as well as other electronic equipment and electrical wires. If you are concerned about this, install your inverter in a place away from people. Avoid locations in which people will be spending a lot of time — for example, don't install the inverter on the other side of a wall from your bedroom or office.

While we are developing a checklist of features to consider when purchasing an inverter, be sure to add ease of programming. Dan's first Trace inverter (DR2424, modified square wave) was a dream when it came to programming: all of the controls were manually operated dials. To change a setting, all he had to do was turn the dial. His new Trace PS2524, a sine wave inverter, works wonderfully (although noisily) in all respects, except when it comes to programming. The digital programming is extremely complicated and the instructions are difficult to follow.

Our advice is to find out in advance how easy it is to change settings, and don't rely on the opinions of salespeople or renewable energy geeks that can recite pi to the 20th decimal place. Ask friends or dealers/installers for their opinions, but also ask them to show you. You may even want to spend some time with the manual (often available online) to see if it makes sense *before* you buy an inverter.

Another feature to look for in a battery-based inverter is power consumption under search mode. What's that?

The search mode is an operation that allows a battery-based inverter to go to sleep, that is, shut down almost entirely in the absence of active loads. The search mode saves energy (inverters may consume 10-30 watts or more from the battery when in the "on" mode).

Although the inverter is sleeping when it is in search mode, it's sleeping with one eye slightly open. That is, it's on the alert should someone switch on a light or an appliance. It's able to do this by sending out tiny pulses of electricity approximately every second. They travel through the electrical wires of a home searching for an active load. When an appliance or light is turned on, the inverter senses the load and quickly snaps into action, powering up and feeding AC electricity to the device.

The search mode is handy in houses in which phantom loads have been eliminated. (As you may recall, a phantom load is a device that continues to draw a small electrical current when off.) According to the Department of Energy, phantom loads, on average, account for about 5% of a home's annual electrical consumption. Eliminating phantom loads saves a small amount of electrical energy in an ordinary home. In a home powered by renewable energy, however, it saves even more because supplying phantom loads 24 hours a day requires an active inverter. An inverter may consume 10 to 30 watts when operating at low capacity. Servicing a 12-watt phantom load, therefore, requires an additional 10 to 30-watt investment in energy in the inverter.

Energy savings created by eliminating phantom loads and continuous inverter operation does have its downsides. For example, automatic garage door openers may have to be turned off for the inverter to go into the search mode. Dan turns off his garage circuit at the main service panel. When he arrives home at night and the circuit's been shut off, he has to switch it back on. Another problem occurs with electronic devices that require tiny amounts of electricity to operate, like cell phone chargers. When left plugged in, many cell phone chargers draw enough power to cause an inverter to turn on. Once the inverter starts, however, the device doesn't draw enough power to keep the inverter going. As

a result, the inverter switches on and off, *ad infinitum*. Dan found the same with his portable stereo.

Because of these problems, many people simply turn the search mode off so that the inverter runs 24 hours a day. Another option is to set the search mode sensitivity up, so it turns on at a higher wattage. However, most modern homes have at least one always-on load, for example, hard-wired smoke detectors or garage door openers that require the continuous operation of the inverter.

Computer Interfacing

Some of the more sophisticated — and most expensive — inverters can be connected to computers for monitoring and reprogramming (to change settings). Such inverters require a computer interface and software. Although this feature may be desirable to some, it's not essential. Inverters can be reprogrammed without a computer.

Stackability

Finally, when buying a battery-based inverter, you may also want to select one that can be stacked — that is, an inverter that can be connected to a second or third inverter of the same kind in parallel or series. Stacking permits homeowners to increase system output — to produce more electricity in case demands increase over time. When wired in parallel, the inverters amp output increases. When wired in series, the voltage increases. A couple of inverter manufacturers such as OutBack and Xantrex produce power panels, easily mounted assemblies that house two or more inverters for stacking. Power panels also contain a number of other components needed in a renewable energy system, such as charge controllers, DC disconnects, and meters. Figure 6-15 shows a power panel by OutBack that contains two inverters. Xantrex's new XW inverters come with similar components.

Stackability is important because many homes require 240-volt electricity to operate appliances such as electric clothes dryers, electric stoves, central air conditioning, or electric resistance heat. As a general rule, we recommend that you avoid such appliances, especially if you're considering an off-grid system. That's not because a PV system can't be designed to power these loads, but rather because they tend to use lots of electricity and you'll need to install a much larger — and much more costly — PV system to power them. Well designed, energy-efficient homes can usually avoid using 240 VAC, which also simplifies home and system wiring. A frequent exception is a deep well pump, which may require 240 volts, but in most cases, a high efficiency 120-volt AC pump, or even a DC pump, can do the job.

Fig. 6-15: Power Panel. Outback's Power Panel and other similar power panels contain most, if not all, of the electronic components required for a successful installation. They simplify installation and reduce installation costs.

If you must have 240-volt AC electricity, purchase an inverter that is designed to be wired in series to produce 240 VAC, such as the OutBack FX-series inverters. Another option is to purchase an inverter that produces 120- and 240-volt electricity, such as Magnum Energy's MS-AE 120/240V series inverter/charger or one of Xantrex's new XW series inverters. Yet another option is to install a step-up transformer such as the Xantrex/ Trace T-240. This unit takes 120-volt AC electricity from an inverter and steps it up to 240-volt AC electricity. Or, you can simply install a dedicated 240-volt output inverter for that load.

Conclusion

If you are hiring a professional to design and install a PV system, you won't have to worry about buying the proper inverter. When installing your own system, be sure that the supplier takes the time to determine which inverter is right for you. Ask lots of questions.

A good inverter is key to the success of a renewable energy system, so shop carefully. Size it appropriately. Be sure to consider future electrical needs. But don't forget that you can trim electrical consumption by installing efficient electronic devices and appliances. Efficiency is always cheaper than adding more capacity. When shopping, select the features you want and buy the best inverter you can afford. Although modified square wave inverters work for most applications, it is best to purchase a sine wave inverter.

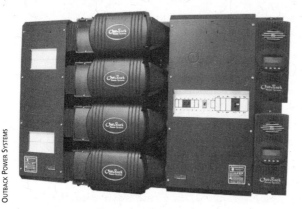

CHAPTER 7

BATTERIES, CHARGE CONTROLLERS AND GEN-SETS

Batteries are the unsung heroes of off-grid PV systems. Even though PV modules receive most of the attention, work hard, and endure extreme weather, the batteries of off-grid systems operate day in and day out, 365 days a year for up to ten years, give or take a little, provided you treat them well. Although batteries in off-grid systems do indeed work hard, these brutes of the renewable energy field are not as brutish as you might think. They require considerable pampering. You can't simply plop them in a battery room, wire them into your system, and then go about your business expecting your batteries to perform at 100% of their capacity for the lifetime of your system. You'll need to watch over them carefully and tend to their needs. If you don't, they'll die young, many years before their time.

Because batteries are so important to off-grid renewable energy systems and require so much attention, it is essential that readers who

are thinking about installing an off-grid system or a grid-connected system with battery backup understand a little bit about batteries. Doing so will help you get the most from them. This chapter will also help you develop a solid understanding of two additional components required in PV battery-based systems: charge controllers and generators. They are vital to the health and vitality of a battery bank.

If you are only interested in a batteryless grid-connected system, you don't need to read this chapter. Feel free to skip on to Chapter 8.

Understanding Lead-Acid Batteries

Batteries are a mystery to many people. How they work, why they fail, and what makes one type different from another are topics that can boggle the mind. Fortunately, the batteries used in most off-grid renewable energy systems are

The Life of a Battery

Off-grid systems require batteries. Without them, there would be no electricity during the periods when PV systems are idle. Batteries carry off-gridders through on a day-to-day basis, ensuring a steady supply of electricity. When the Sun is not shining or when demand exceeds output of the PV array, batteries are there to provide the electricity needed to keep homes and offices running smoothly. When surplus electricity is available, the batteries store it for later use.

While batteries in an off-grid system work day and night, batteries in a grid-connected system live a life of leisure. Their main function is to store electrical energy in case of an emergency. In between power outages, however, the batteries lounge in the comfort of the battery room, filled to capacity with electricity, which is as good as it gets for a battery. Their sole purpose is to remain fully charged in case they are called into duty. Other than creating a sense of security in between occasional power outages, they serve no function.

As a general rule, the largest (and most costly) battery banks are required for off-grid systems — for example, PV or wind electric systems. Wind/PV hybrid systems require slightly smaller battery banks, for reasons explained in Chapter 5. Grid-connected PV systems with battery backup generally require even smaller battery banks — typically just enough to get a family or business by for a few hours or a few days should grid power fail.

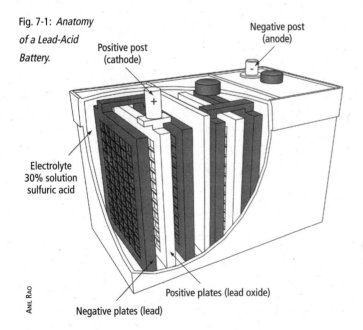

Fig. 7-1: *Anatomy of a Lead-Acid Battery.*

Negative post (anode)

Positive post (cathode)

Electrolyte 30% solution sulfuric acid

Positive plates (lead oxide)

Negative plates (lead)

ANIL RAO

pretty much the same: they're deep-cycle, flooded lead-acid batteries.

Flooded lead-acid batteries for renewable energy systems are the ultimate in rechargeable batteries. They can be charged and discharged (cycled) a thousand or more times before they wear out.

Lead-acid batteries used in most renewable energy systems contain three cells — that is, three distinct compartments, each of which produces about two volts. Inside the battery case, the individual cells are electrically connected (wired in series). As a result, they collectively produce 6-volt electricity.

Inside each cell in a flooded lead-acid battery is a series of thick, parallel lead plates, as shown in Figure 7-1. Each cell is filled with

$$Pb(s) + HSO_4{}^-(aq) \longrightarrow PbSO_4(s) + H^+(aq) + 2e^- \qquad \text{negative plates}$$

$$PbO_2(s) + HSO_4{}^-(aq) + 3H^+(aq) + 2e^- \longrightarrow PbSO_4(s) + 2H_2O(1) \quad \text{positive plates}$$

Fig. 7-2: *Chemical Reactions in a Lead-Acid Battery. The chemical reactions occurring at the positive and negative plates are shown here. Note that these reactions are reversible. Lead sulfate forms on both the positive and negative plates during discharge.*

battery fluid, hence the term "flooded." Battery acid is distilled water and sulfuric acid. A partition wall separates each cell, so that fluid cannot flow from one cell to the next. The cells are encased within a heavy-duty plastic case.

As illustrated in Figure 7-1, two types of plates are found inside a battery: positive and negative. The positive plates are connected internally to the positive post or terminal. The negative plates are connected internally to the negative post. The posts allow electricity to flow into and out of the battery.

As shown in Figure 7-1, the positive plates of lead-acid batteries are made from lead dioxide (PbO_2). The negative plates are made from pure lead. Sulfuric acid fills the spaces between the plates and is referred to as the *electrolyte*. It plays an extremely important role in batteries, and will be discussed shortly.

How Lead-Acid Batteries Work

Although battery chemistry can be a bit daunting to those who have never studied the subject or had it explained well, it's important that all users understand a little about the reactions.

Like all other types of batteries, lead-acid batteries convert electrical energy into chemical energy when charging. When discharging, that is, giving off electricity, chemical energy is converted back into electricity. How does this occur?

As shown in the top reaction in Figure 7-2, when electricity is drawn from a lead-acid battery, sulfuric acid reacts chemically with the lead of the negative plates. This chemical reaction yields free electrons, tiny negatively charged particles. They flow out of the battery via the positive terminal through the battery cable. During this chemical reaction, lead reacts with sulfate ions, forming lead sulfate. As a result, tiny lead sulfate crystals form on the surface of the negative plates when the battery is discharging.

The second equation in Figure 7-2 shows the chemical reaction that takes place at the positive plates when a battery is discharging; sulfuric acid reacts with the lead dioxide of the positive plates, forming tiny lead sulfate crystals on them as well.

Discharging a battery not only creates lead sulfate crystals, it converts some of the sulfuric acid into water. However, as you shall soon see, the chemical reactions occurring at the positive and negative plates when a battery is discharging are reversible. That is to say, when a battery is charged (when electricity is delivered to the battery), the chemical reactions at the positive and negative plates run in reverse. As a result, lead sulfate crystals on the surface of the positive and negative plates are broken down and sulfuric acid is regenerated. The positive and negative plates are restored. Electrons are "stored" in the chemicals in the system.

Although the chemistry of lead-acid batteries is a bit complicated, it is important to remember that this system — and other battery systems as well — work because electrons can be stored in the chemicals within the battery when a battery is charged. The stored electrons can be drawn out by reversing the chemical reactions. Through this reversible chemical reaction, the battery stores electricity, making it available when needed.

Will Any Type of Lead-Acid Battery Work?

Lead-acid batteries are used in a wide range of applications. For example, there's a 12-volt flooded lead-acid battery under the hood of your automobile. Trucks and buses have similar batteries, only larger. In all of these vehicles, batteries provide the electricity required to start the engine and operate lights, the ignition system and accessories like a clock or radio when the engine's not running. These are known as starting, lighting and ignition, or *SLI, batteries*.

Lead-acid batteries are also used in a wide assortment of electric vehicles, including forklifts, golf carts and electric lawn mowers.

A Word of Warning

Sulfuric acid is a *very* strong acid. In fact, it is one of the strongest acids known to science. In flooded lead-acid batteries, sulfuric acid is diluted to 30%. Although diluted, it is still a chemical to treat with great respect — it can burn your skin and eyes and eat through clothing like a ravenous moth.

They are even used in emergency standby power systems for businesses, providing backup power in case the electrical grid goes down. Lead-acid batteries are also used in RVs, sailboats, yachts and powerboats, although sealed batteries are typically used in these applications so they won't spill acid when the boat or RV rocks. (More on sealed batteries shortly.)

Lead-acid batteries come in many varieties, each one designed for a specific application. Car batteries, for example, are designed for use in cars, light trucks and vans. Their thin, porous plates offer lots of surface area to the electrolyte to enable the battery to crank out the large number of amps needed to start a car or truck. These thin plates can easily be damaged by deep discharge. Marine batteries are similar but are optimized for use in boats — for starting engines and providing small amounts of electricity to power electrical equipment like radios and GPS units.

While most lead-acid batteries — even car batteries — can be installed in a renewable energy system, for optimum performance and years of trouble-free service, we recommend using batteries designed for use in renewable energy systems: deep-cycle, flooded lead-acid batteries (Figure 7-3). They can be safely deep discharged over and over again with relative impunity, for reasons explained shortly.

Even though batteries designed for deep discharge are considerably more robust than car and truck batteries, they are not invincible. If a battery is discharged too deeply — more than 80% of its capacity — it can be permanently damaged. Fortunately, controls in a well-designed renewable energy system prevent this from occurring.

For optimum performance, deep-cycle batteries also need to be recharged fairly soon after undergoing a deep discharge. As noted earlier, discharging a battery produces lead sulfate crystals on the plates. If the batteries are left in a state of discharge for an extended period, the lead sulfate crystals begin to grow. Small crystals become large. These crystals reduce the ability of the battery to store electricity. Batteries take progressively less charge, and have less to give back. Large crystals are also not completely converted back to lead or lead dioxide during subsequent charge. Over time, then, entire cells may die, effectively ending that battery's useful life.

To avoid lead sulfate crystal growth, batteries should be promptly recharged and also periodically *equalized*. This process, discussed shortly, helps rejuvenate batteries by driving lead sulfate off the plates. It also equalizes the voltage of the cells in a battery bank.

So, let there be no lingering doubts in your mind: car and truck batteries are no match for powerful lead-acid batteries designed for the deep cycling that typically occurs in renewable energy systems. Do not forget: *For optimum long-term performance, batteries need to be recharged promptly after deep discharging takes place to ensure long life.* Deep-cycle, flooded lead-acid batteries like L16s,[1] if they are promptly recharged after deep discharging, routinely equalized, and housed in a warm location, should last for seven to ten years, maybe even longer. If you fail to follow this routine maintenance, you can count on having to get your checkbook out.

Can I Use a Forklift, Golf Cart or Marine Battery?

Forklifts require high capacity, deep-discharge batteries. These leviathans can be used in renewable energy systems because they are designed for a long life and operate under fairly demanding conditions. In fact, they can withstand 1,000 to 2,000 deep discharges — more than many other deep-cycle batteries used in battery-based renewable energy — making them ideally suited for renewable energy systems.

Although forklift batteries function very well in such instances, they are rather heavy and expensive. If you can acquire them at a decent price, you might want to use them, especially if the size of your system warrants so large a battery.

Golf cart batteries can also work. Like forklift batteries, golf cart batteries are designed

Fig. 7-3: *Lead-acid batteries are commonly used in renewable energy systems. Be sure to purchase deep-cycle batteries, keep them in a warm, safe place, recharge quickly after deep discharges, and equalize them periodically.*

for deep discharge. Moreover, they typically cost a lot less than larger deep-cycle batteries. While the lower cost may be appealing, golf cart batteries don't store as much electricity as other deep-cycle batteries and don't last as long as the alternatives. They may last only five to seven years, if well cared for. Shorter lifespan means more frequent replacement, which means higher long-term costs and more hassle.

Although golf cart batteries don't last as long as the heavier-duty, deep-cycle batteries like L16s, they are inexpensive and can be used in PV systems. In fact, some renewable energy system installers swear by them. Others recommend them for first-time users, who are bound to abuse their first set of batteries. They're a training set.

Although marine batteries are advertised as deep-cycle or "marine deep-cycle" batteries, they are really a compromise between a car starting battery and a deep-cycle battery. Their thinner plates just aren't up to the demands of a renewable energy system. They will not last as long in deep-cycle service as true deep-cycle batteries.

What About Used Batteries?

Another option for cost-conscious home and business owners is a used battery. Although used batteries can sometimes be purchased for pennies on the dollar, and may seem like a bargain, for the most part, they're not worth it. As a buyer, you have no idea how well — or more likely, how poorly — they've been treated. How old are they? Have they been deeply discharged many times? Have they been left in a state of deep discharge for long periods? Have they been filled with tap water rather than distilled water? Have the plates been exposed to air due to poor maintenance?

Many used batteries are discarded because they've failed or have experienced a serious decline in function. As we've said, with age comes decreased capacity and also decreased efficiency. If a used battery is 90% efficient, you have to put 110 amp hours into the battery for every 100 amp hours you take out. If the battery is only 50% efficient, you have to put 200 amp hours into it for every 100 amp hours you draw out. Put another way, you'd have to have a PV array that is twice the size (and cost) because you saved money buying old, used batteries.

Although there are exceptions, most people who've purchased used flooded lead-acid batteries have been disappointed.

Sealed Batteries

While most installers of off-grid PV systems recommend flooded lead-acid batteries, they

Don't Skimp on Batteries

When shopping for batteries for a renewable energy system, we recommend that you buy high quality deep-cycle batteries. Although you might be able to save some money by purchasing cheaper alternatives, frequent replacement is a pain in the neck. Batteries are heavy and it takes quite a lot of time to disconnect old batteries and rewire new ones. Bottom line: the longer a battery will last — because it's the right battery for the job and it's well made and well cared for — the better.

install "sealed" lead-acid batteries in grid-connected systems with battery backup. Sealed batteries are also known as *captive electrolyte batteries*. They are filled with electrolyte at the factory, charged, and then permanently sealed. Because they are sealed, they are easy to handle and ship without fear of spillage. They won't even leak if the battery casing is cracked open, and they can be installed in any orientation — even on their sides. Unlike flooded lead-acid batteries, which require periodic filling with distilled water, sealed batteries never need watering. In fact, you *can't* add distilled water.

Two types of "sealed" batteries are available: *absorbed glass mat* (AGM) batteries and *gel cell batteries*. In absorbed glass mat batteries, thin absorbent fiberglass mats are placed between the lead plates. The mat consists of a network of tiny pores that immobilize the battery acid. The mat also creates tiny pockets that capture hydrogen and oxygen gases given off by the lead plates during charging. The gases recombine in these pockets, reforming water. Unlike a flooded lead-acid battery, the gases can't escape. That's why sealed AGM batteries never need watering.

In gel batteries, the sulfuric acid electrolyte is converted to a substance much like hardened Jell-O by the addition of a small amount of silica gel. The gel-like substance fills the spaces between the lead plates.

Sealed batteries are often referred to as "maintenance-free" batteries because fluid levels never need to be checked and because the batteries never need to be filled with distilled water. They also never need to be (and should

Sealed Batteries — A Misnomer?

The truth be known, sealed batteries are not totally sealed. Each battery contains a pressure release valve that allows gases and fluid to escape if a battery is accidentally overcharged. The valve keeps the battery from exploding. Once the valve has blown, though, the battery will very likely need to be retired.

not be!) equalized, a process discussed later in the chapter. Eliminating routine maintenance saves a lot of time and energy. It makes sealed batteries a good choice for off-grid systems in remote locations where routine maintenance is problematic — for example, backwoods cabins or cottages that are only occasionally occupied. Sealed batteries are also used on sailboats and RVs where the rocking motion could spill the sulfuric acid contained in flooded lead-acid batteries. In these applications, space is limited and batteries are frequently crammed into out-of-the way locations. Sealed batteries are also used in battery-based grid-connected systems.

Sealed batteries offer several additional advantages over flooded lead-acid batteries. One advantage is that they charge faster. Sealed batteries also release no explosive gases, so there's no need to vent battery rooms or battery boxes where they're stored. In addition, sealed batteries are much more tolerant of low temperatures. They can even handle occasional freezing, although this is never recommended. Sealed batteries also experience a lower rate of self-discharge. That is, they discharge more slowly than flooded lead-acid batteries when not in use. (All batteries self-discharge when not in use.)

Sealed batteries do have a few significant disadvantages. One of them is that they are much more expensive than flooded lead-acid batteries. Another is that they store less electricity than flooded lead-acid batteries of the same size. Sealed batteries have a shorter lifespan than flooded lead-acid batteries. In addition, they can't be rejuvenated like a standard lead-acid battery if left in a state of deep discharge for an extended period. In other words, large lead sulfate crystals that form on the plates of a sealed battery can't be removed through equalization. Equalization, while safe in flooded lead-acid batteries, results in excessive pressure buildup inside a sealed battery. Pressure would be vented through the pressure release valve on the sealed battery, causing electrolyte loss that could destroy or seriously decrease the storage capacity of the battery. Finally, sealed batteries must be charged at a lower voltage setting than flooded lead-acid batteries.[2] Because of these problems, most installers working with battery-based systems recommend flooded lead-acid batteries.

Wiring a Battery Bank

Batteries are connected together to produce the voltage and storage capacity required by a renewable energy system. Small renewable energy systems — for example, those found in RVs, boats and cabins — are typically wired to produce 12-volt electricity. Many of these applications run entirely off 12-volt DC electricity. Systems in off-grid homes and businesses are typically wired to produce 24- or 48-volt DC electricity. The low-voltage DC electricity, however, is converted to AC electricity by the inverter, which also boosts the voltage to the 120 and 240 volts commonly used in homes and businesses.

How to wire a battery bank is beyond the scope of this book, but you should know some of the basics. One of them is that batteries can be wired in series or parallel. Wiring batteries in series means that the positive terminal of one battery is connected to the negative terminal of the next one and so on, as shown in Figure 7-4. As you may recall from high school physics, wiring batteries in series increases the voltage. Two 1.5-volt batteries in series in a flashlight produce 3-volt electricity. In a PV system, four 6-volt lead-acid batteries wired in series produce 24-volt DC electricity. Eight 6-volt batteries wired in series produce 48-volt DC electricity (Figure 7-4). A group of batteries wired in series is called a *string*.

While wiring in series boosts the voltage, wiring in parallel increases the amp-hour storage capacity of a battery bank — the number of amp-hours of electricity a battery bank can store. (One amp-hour is one amp of current flowing for one hour — either into or out of a battery bank.) To boost the amp-hour capacity of a renewable energy system, installers typically include two or more strings of batteries in parallel, as shown in Figure 7-5. As you can see from the diagram, each string is wired in series; however, the individual strings are connected in parallel. As shown in Figure

7-5, in parallel wiring the positive ends of each string are connected, as are the negative ends. Wiring in parallel increases the amp-hour rating.[3]

Sizing a Battery Bank

Properly sizing a battery bank is critical to designing a reliable off-grid PV system. The principal goal when sizing a battery bank is to install a sufficient number of batteries to carry your household or business through periods when the Sun is not available.

In off-grid systems, the easiest way to size a battery bank is to calculate how much electricity you use in a day. Because electrical consumption varies from month to month, it is best to use a daily rate from the month of greatest consumption. In many locations, this

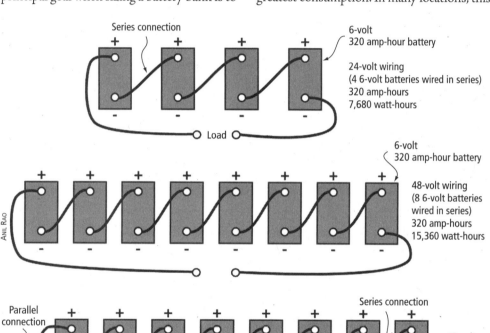

Fig. 7-4: *Batteries are wired in a string in series to boost the voltage. Several strings can be wired in parallel to boost amp-hour storage. The battery banks here are wired for 24 and 48 volt electricity.*

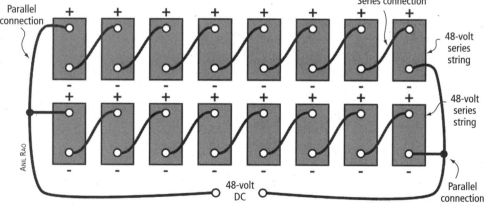

Fig. 7-5: *In this diagram, two 48-volt strings are wired in parallel to boost the number of kilowatt-hours of electricity that are stored and available to a homeowner or business.*

More on Sizing a Battery Bank

To understand how a battery bank is sized, let's look at an example. Let's suppose that you and your family consume 4 kWh (4,000 watt-hours) of electricity a day on average during the most energy-intensive month of the year. If you need five days of battery backup, you will need 20 kWh of storage capacity (5 days x 4 kWh).

However, it's best not to discharge batteries below the 50% mark because deeply discharging batteries reduces their life. So, if you set 50% as your discharge goal, you'll need 40 kWh — or 40,000 watt-hours — of storage capacity.

To determine how many batteries you need to provide this amount of electricity, first find out how much electricity can be stored in the batteries you are thinking about buying. Battery storage capacity is rated in amp-hours. For instance, a battery might be rated at 300 amp-hours. But how do you know how many amp-hours of storage capacity you need when all you know is that you need 40 kWh of storage capacity?

The answer is to convert kilowatt-hours to amp-hours. To make this conversion, installers simply divide watt-hours by the voltage of the system. (Remember the equation: watts = amps x volts. To solve for amp-hours, divide watt-hours by volts). If you are installing a 24-volt system, 40,000 watt-hours divided by 24 volts results in 1,666 amp-hours. That's the required battery storage capacity.

The next task is to see how many batteries you will need to store this much electricity. You do that by checking the manufacturer's specifications on their batteries. The Trojan L-16Hs that Dan uses in his hybrid solar electric/wind system are 6-volt batteries, each with a rated capacity of 420 amp-hours. To determine how many of these batteries you'd need, it would seem that all you'd need to do is divide 1,666 amp-hours of storage capacity by 420 amp-hours of battery storage. The result would be four batteries...

Well, not quite.

If you recall from previous discussions, wiring four 6-volt batteries in series increases the voltage to 24 volts, but does not change the amp-hour storage capacity. That is to say, a string of four 6-volt batteries, each with a 420 amp-hour capacity, will provide 24-volt DC electricity, but will store only 420 amp-hours of electricity. To create 1,666 amp-hours of storage, you'd need four strings, each one of which contained four 6-volt batteries. Each string would yield 420 amp-hours of electricity. Combined in parallel, the four strings would provide the 1,600 plus amp-hours required to meet your needs.[4]

A simpler way to run these calculations is to convert amp-hour storage capacity of batteries to kilowatt-hours at the outset. A 420 amp-hour 6-volt battery, for example, contains 2,520 watt-hours or 2.5 kWh (watt-hours = amp-hours x volts). Four 6-volt batteries, therefore, contain 10,080 watt-hours or 10.08 kWh. To store 40,000 watt-hours, you'd need four strings, each with four 6-volt batteries.

As you can imagine, battery storage can become quite expensive. In the example we just used, 16 batteries could cost $4,500 (not including installation costs!).

occurs during the dead of winter, the least sunny time of year.

Once you determine how many kilowatt-hours of electricity your home or office will consume on average during a typical day during your most energy-intensive month, you must make an educated guess as to the number of sunless days that occur during that time. That is, you must determine how long your renewable energy system will be sidelined by a lack of sunlight. In most cases, renewable energy designers plan a three-day reserve period. That is, they plan for three days of no sun. (Bear in mind that even on cloudy days, solar electric modules continue to produce electricity, albeit at a significantly reduced rate.) Battery banks are therefore sized to meet the electrical demand for this period. Longer reserve periods — five days or more — may be required for some areas. Shorter periods are generally required for hybrid systems. Shorter reserve periods may also be used for those who are willing to run a backup fossil-fuel generator during cloudy periods. Then, the main question becomes: "How frequently do I want to operate my generator?" If you're interested in learning more about battery banks, read the accompanying sidebar, "More on Sizing a Battery Bank."

Reducing the Size of Battery Banks

Because batteries are expensive, require periodic maintenance, and take up a lot of room, many off-grid homeowners install hybrid systems — systems that often combine solar electricity and a wind generator. This allows them to reduce the size of their battery banks. Most homeowners install a gen-set — a gasoline, diesel, propane or natural gas generator — to provide additional backup power. Or they can increase the size of their array. Because PV modules will very likely outlast four or five battery banks, the investment in additional generating capacity may be well worth it in the long run. In addition, PV systems require very little maintenance.

Home Power's Ian Woofenden points out, however, "although I love the theory of investing in RE capacity instead of lead (in batteries), there is a limit to its potential." In cloudy regions, PVs may not produce enough electricity to make a huge difference during winter months. In such instances, batteries and gen-sets may be a preferred option.

Fortunately, the National Renewable Energy Laboratory offers online assistance to people trying to figure out the best combinations. The program, called *HOMER*, is designed to optimize hybrid power systems (to find it, visit analysis.nrel.gov/homer/). "This is a great tool for optimizing the size of various components in a hybrid power system," notes National Renewable Energy Laboratory's small wind expert Jim Green. "It typically shows that adding a backup generator will reduce system cost and that optimum battery bank size may be on the order of 8 to 12 hours of storage, much less than 3 days, which is typically used." He adds, "Of course, the optimization point will change over time as the price of propane and natural gas go up. Even so, the generator run-time can be quite low … This model [HOMER] will be more than most homeowners will want to tackle, but some will be able and willing to use it. Equipment dealers find it to be a useful tool."

Battery Maintenance and Safety

Now that you understand lead-acid batteries, your options, and a little about wiring and sizing a battery bank, it is time to turn to the equally important topics of battery care, maintenance and safety. Battery care and maintenance are vital to the long-term success of a renewable energy system. Proper maintenance increases the service life of a battery. Because batteries are expensive, longer service life results in lower operating costs over the long haul. Put another way, the longer your batteries last, the cheaper your electricity will be. In this section, we'll examine battery care, maintenance and a few safety issues.

Keep Them Warm

Batteries may be the workhorses of a renewable energy system, but they like to be kept warm — and full. Cold conditions dramatically reduce their capacity — the amount of electricity they'll store (Figure 7-6). That's because low temperatures slow down the chemical reactions in batteries, reducing electrical storage.

While low temperatures reduce battery efficiency, higher temperatures result in an increase in outgassing. Outgassing is the release of explosive hydrogen and oxygen gas, which also results in reduced battery fluid levels. Higher temperatures also lead to higher rates of self-discharge. (Self-discharge is the loss of charge over time when a battery sits idle. As a rule, older batteries lose charge faster than new batteries.) Outgassing is not a trivial matter. Despite what you might think, "Arizona is a harsher climate for car batteries than Minnesota," notes Jim Green.

For optimal function, batteries should be kept at around 75 to 80 degrees — about the same temperatures that we like to live in. In this range, batteries will accept and deliver a lot more electricity. Guaranteed!

If you can't ensure this narrow temperature range, shoot for a range between 50 and 80 degrees Fahrenheit. Rarely should batteries fall below 40 degrees or exceed 100 degrees Fahrenheit.

Ideally, batteries should be stored in a separate battery room or a battery box inside a conditioned space at the optimum temperature. The best place for batteries is a heated space such as a shop, utility room or garage. We don't encourage people to house their batteries inside their homes, although Dan houses his in a well-sealed and vented battery closet in his utility room, which works well.

Fig. 7-6: *Battery Performance vs. Temperature. Batteries function optimally above 50° F. Battery capacity decreases rather dramatically as temperatures in the battery room fall.*

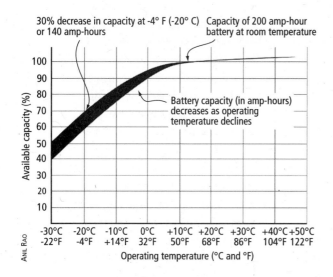

30% decrease in capacity at -4° F (-20° C) or 140 amp-hours   Capacity of 200 amp-hour battery at room temperature

Battery capacity (in amp-hours) decreases as operating temperature declines

Available capacity (%)

Operating temperature (°C and °F)

ANIL RAO

Whatever you do, don't store batteries in a cold garage, barn or shed. Besides delivering less electricity, they won't last as long. Expect very short service lives when batteries are kept cold. If the batteries are in a low state of charge, they can freeze. Freezing can cause the cases of flooded lead-acid batteries to crack. This allows electrolyte to leak out, creating a dangerous mess in the battery room. Deeply discharged batteries are especially prone to freezing because the sulfuric acid has reacted with lead and lead dioxide to form lead sulfate, leaving a higher concentration of water, which makes the battery more prone to freezing.

Batteries should not be stored on concrete floors.[5] Cold floors cool them down and reduce their capacity. Always raise batteries off the floor so they stay warmer.

If you must store batteries outside your home, be sure to heat and cool the building. Heating and cooling a shed, garage or barn can be quite expensive and wasteful of valuable fossil fuel energy. If you must heat a battery room, we recommend retrofitting the storage room for passive solar or building a storage facility that's passively heated and cooled — or perhaps installing a solar hot air collector. Insulate the building to the max and provide south-facing glass to keep it warm — but not too warm — in the winter. (For more on passive solar design, you may want to read about these topics in Dan's book *The Solar House: Passive Heating and Cooling*.)

Ventilating Flooded Lead-Acid Batteries

To ensure safe operation, battery boxes should also be ventilated to the outside to remove potentially explosive hydrogen and oxygen gas released when flooded batteries charge. (This does not apply to the sealed batteries often used in grid-connected systems with battery backup.) A small 2-inch vent made from PVC or ABS pipe is all that's typically required to vent a battery box.

Battery rooms may also require venting, as shown in Figure 7-7. However, hydrogen disperses fairly rapidly and therefore requires very little air movement to prevent accumulation. Natural air movement in a battery room combined with normal air changes that occur in occupied spaces or as a result of heating systems will normally be sufficient to disperse hydrogen gas from battery rooms.

When venting a battery box or confined battery room, be sure that the vent system doesn't cool the space in the winter. Also, be sure that activities that require an open flame or that might produce sparks are never carried out near a battery bank.

As shown in Figure 7-7, proper ventilation requires small air inlets, openings that allow fresh air into the battery box or battery room near floor level. It also requires an air outlet near the top of the box or ceiling to vent hydrogen gas to the outside. Check with local electrical codes to be sure you comply with their requirements, if any.

Battery rooms are generally passively ventilated. However, they can also be actively vented or power vented. Power venting requires a small electric fan that exhausts hydrogen gas in the enclosure while batteries are charging. As nifty as power venting may seem, it's generally not necessary. Hydrogen is an

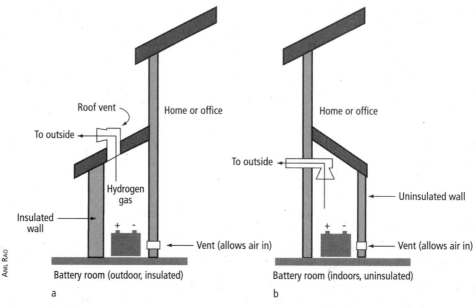

Roof vent
Home or office
To outside
Hydrogen gas
Insulated wall
+ −
Vent (allows air in)
Battery room (outdoor, insulated)
a

Home or office
To outside
Uninsulated wall
+ −
Vent (allows air in)
Battery room (indoors, uninsulated)
b

ANIL RAO

Fig. 7-7: Battery Vent System. Flooded lead-acid batteries require venting to remove hydrogen gas given off during charging — by the PV array and backup generator or grid.

extremely light gas and easily escapes if a room is properly vented. So unless the vent pipe is long and contorted or you live in a cold climate, you probably don't need to power vent a battery bank. If, however, batteries are located in a room used for other purposes, for example, a shop or garage, active venting may be helpful. It reduces the chance of explosions and rids the room of smelly gases produced when batteries are charging. (To learn about power venting from one manufacturer, visit zephyrvent.com.)

Battery Boxes

Large battery banks are often housed in separate, lockable and ventilated rooms. Rather than dedicating an entire room to batteries, however, many homeowners install batteries in sealed and ventilated battery boxes in rooms that serve other purposes. Battery boxes are typically built from plywood. Some individuals build battery boxes with solid wooden lids; others incorporate clear plastic (polycarbonate) lids. Be sure to line the boxes with an acid-resistant liner to contain possible acid spills. Lids should be hinged and sloped to discourage people from storing items on top. Also, be sure to place battery boxes in a warm room. Raise the boxes off cold concrete floors and be sure that the supports are sufficient to hold the weight of the batteries. L16 size batteries weigh about 125 pounds each. Also, be sure to seal battery boxes and vent them to the outside.

Batteries can also be stored in battery boxes made from off-the-shelf plastic tubs. Tubs need to be modified to provide ventilation and should be raised off the floor.

Battery storage boxes can be purchased from commercial outlets. Radiant Solar Tech, at radiantsolartech.com, for example, manufactures sealed plastic battery boxes with removable lids. This company will also custom build boxes for homeowners.

When shopping for battery boxes, note that many commercially available products are designed for sealed batteries, not flooded lead-acid batteries. Before you buy a manufactured battery storage container, be sure that it will work with the batteries you are planning on using in your system. Containers for flooded lead-acid battery banks should provide sufficient clearance to view electrolyte levels and to fill cells when battery fluid levels drop.

Finally, batteries should be located as close to the inverter and other power conditioning equipment as possible. Doing so minimizes voltage drop and power losses and will help reduce system costs.

Keep Kids Out

If young children are likely to be about, battery rooms and battery boxes should be inaccessible and locked. This will prevent kids from coming in contact with the batteries and risking electrical or acid burns.[6]

Managing Charge to Ensure Longer Battery Life

Keeping flooded lead-acid batteries warm and topped off with distilled water ensures a long lifespan and optimum long-term output. Longevity can also be ensured by proper charge management — specifically keeping batteries as fully charged as possible. To understand why, you must understand cell capacity, that

is, the amount of electricity a battery can store, and the effect of discharge rates on cell capacity.

"Cell capacity is the total amount of electricity that can be drawn from a fully charged battery until it is discharged to a specified battery voltage," according to Richard J. Komp, author of *Practical Photovoltaics: Electricity from Solar Cells*. Battery cell capacity is measured in amp-hours. A Trojan L16H battery can cycle approximately 420 amp-hours of electricity. Theoretically, a battery with a 420-amp-hour storage capacity could deliver one amp of electricity over a 420-hour-period or 420 amperes for one hour. In reality, battery capacity — the amount of electricity that can be drawn from a battery — varies with the discharge rate, that is, how fast a battery is discharged. As a rule, the faster a lead-acid battery is discharged, the less you'll get out of it. A lead-acid battery discharged over a 20-hour period, for example, will yield 100% of its rated capacity.[7] Discharge that battery in

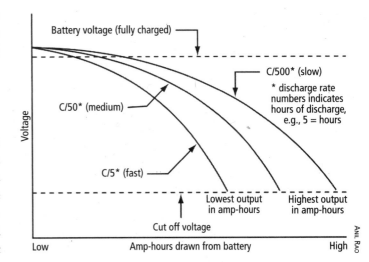

Fig. 7-8: *Discharge Rate vs. Amp-Hours Delivered. The rate at which a battery bank is discharged determines how much energy can be obtained. The slower the rate of discharge, the more energy.*

an hour and a half, and it will deliver only 75% of its rated capacity. Figure 7-8 illustrates the energy removed from batteries at various discharge rates.

To standardize the industry, most batteries are rated at a specific discharge rate, usually 20 hours (which is what the 420-amp-hour capacity in the example above is based on).

Like many devices, lead-acid batteries last longer the less you use them. More specifically, the fewer times a battery is deeply discharged, the longer it will last. As illustrated in Figure 7-9, a lead-acid battery that's regularly discharged to 50% will last for slightly more than 600 cycles, if recharged after each deep discharge. If regularly discharged no more than 25% of its rated capacity — and recharged after each discharge — the battery should last about 1,500 cycles. If the battery is regularly discharged only 10% of its capacity, it will last for 3,600 cycles. Even though deep-cycle batteries can handle deep discharges, shallow cycling makes them last longer.

As Ian Woofenden points, out, however, this topic — like so many issues in life — is a

bit more complicated. While deep discharging reduces the lifespan of a battery, renewable energy users are more concerned with the cost of the battery per watt-hour cycled. In other words, what we want from batteries is not simply to "last a long time" but to cycle a lot of energy.

The cost of a battery per cycled kWh, says Ian, provides a better measure than total years of service. Theoretically, you'll get the most bang for your buck by cycling in the 40 to 60% discharge range. This goes against the "shallow cycling makes batteries last longer" idea. (After all, if you double the size of a battery bank and reduce the discharge cycle by half, will it last twice as long? If so, you'd only break even on that larger investment.)

In off-grid renewable energy systems, minimizing very deep discharging ensures longer battery life. However, battery life also depends on frequent recharging of batteries after deep discharges. This drives lead sulfate off the plates. Whatever you do, never leave batteries at a low state of charge for a long time. Lead sulfate crystals will enlarge and could

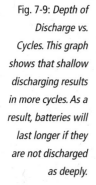

Fig. 7-9: *Depth of Discharge vs. Cycles. This graph shows that shallow discharging results in more cycles. As a result, batteries will last longer if they are not discharged as deeply.*

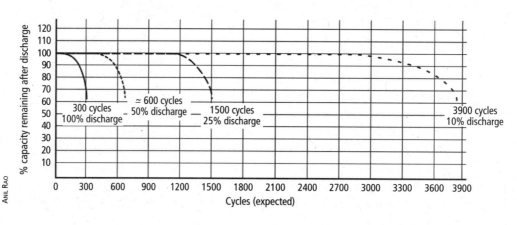

Anil Rao

damage your batteries. A good rule to follow is to be sure that the battery bank is fully charged at least once a week (and fully charged does not mean mostly charged!). Think of it this way, if you don't top them off, you start losing the top, starting a downhill slide.

Protecting batteries from deep discharge is easier said than done. As Ian points out, "in real life you often don't have such close control of how deeply your batteries cycle." If your system is small and you don't pay much attention to electrical use, you'll very likely overshoot the 40 to 60% mark time and time again. If you monitor your electrical usage more carefully, and charge batteries shortly after periods of deep discharge, you are more likely to achieve this goal.

One way of reducing deep discharge is to conserve energy and use electricity as efficiently as possible. Conserving energy means not leaving lights or electronic devices running when they're not in use — all the stuff your parents told you when you were a kid. It also means installing energy-efficient lighting, appliances, electronics, and so on — the ideas many energy conservation experts have been suggesting for decades. In addition, it means ridding your home or business of phantom loads.

Conserving energy and making your home or business as energy-efficient is only half the battle, however. In an off-grid system, you may also have to adjust electrical use according to the state of charge of your batteries. In other words, you may need to cut back on electrical usage when batteries are more deeply discharged and shift your use of electricity to times when the batteries are more fully charged. You may, for instance, run your washing machine and microwave on a bright sunny day when your batteries are full, but hold off during cloudy periods when the batteries are running low — unless you want to run a backup generator. Learning how to monitor the state of charge in your batteries is a skill you can learn, like any other.

After living with solar electricity and wind energy since June 1996, Dan watches the weather. If he has experienced a couple of cloudy, windless days, he knows that he has to curtail his electrical consumption. He may cook on the gas-powered range, as opposed to the microwave. He may hold off on laundry until the Sun comes out again.

Dan monitors his system by checking the battery voltage on the meter in his power center at the *end* of the day (when the sun's not shining, the wind's not blowing, and there's little, if any, electricity being drawn from the battery bank). The voltage reading gives him a clue about the state of charge of his batteries. If the batteries are truly "at rest," low voltage readings mean the batteries are in a low state of charge. Higher readings indicate that the batteries are full.

Voltage is a pretty crude way of monitoring battery capacity, but it is better than nothing, if you know how to interpret it. Just remember, voltage and battery state of charge are not directly related unless the batteries are at rest. At rest means they are not being charged or discharged or have not been recently charged or discharged. A battery is only considered to be at rest when there has been no flow of

Understanding Battery Charging

Batteries are charged by the PV array via the charge controller when a PV system is producing current. The charge controller delivers the array's output (DC electricity) to the battery bank in a controlled manner, terminating the flow of electricity if the battery's voltage climbs to a certain point to prevent overcharging.

Batteries can also be charged by a battery charger located in a modern battery-based inverter. The charger converts 120-volt alternating current electricity from the gen-set (or the utility line in grid-connected systems with battery backup) to direct current electricity. It also decreases the voltage so it matches the voltage of the battery bank. The inverter-based battery charger contains a rectifier (to convert AC to DC) and a step-down transformer to reduce the voltage.

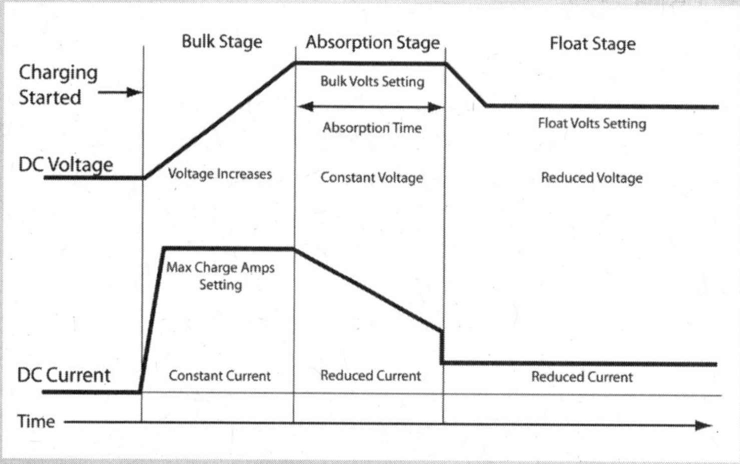

Battery charging occurs in three stages. The first stage of battery charging is known as the *bulk stage*. During this stage, shown in the bottom panel in Figure 7-10, the battery charger (or charge controller when the PV array is charging a battery bank) delivers a constant and relatively high charge (amps). As the charging source "pumps" amps into the battery bank, the battery voltage steadily increases, as shown in the top panel in Figure 7-10. Once the battery voltage reaches a certain point, known as the *voltage regulation set point*, or bulk volts setting, the bulk stage ends. At the end of the bulk stage, the battery is about 80-90% charged.

In the second stage, known as the *absorption stage*, the battery voltage is kept constant and the charge rate (amps) slowly declines. The absorption stage typically lasts about two hours to ensure that a battery is fully charged. (Absorption time is determined by the type of battery and is an adjustable setting.)

At the end of the absorption stage, the charging cycle enters the third and final stage, the *float stage*. During the float stage, the voltage is reduced slightly and held constant at the float volt setting. The current is maintained at a level just a little higher than that which is needed to offset the self-discharge rate of the battery. ☞

Fig. 7-10: *DC Voltage and AC Current During Battery Charging. See text for explanation.*

In an off-grid system with a generator controlled automatically by the inverter, the generator is switched off as soon as the charger enters the float stage. This minimizes generator run time and fuel consumption In a grid-connected system. The float stage, however continues with current supplied by the PV array or the utility grid , depending on the time of day and the programmed settings.

Equalization

Equalization is sometimes referred to as the fourth stage of battery charging. Batteries are usually equalized by external sources, such as gen-sets or wind turbines, because PV arrays are generally unable to supply sufficient current for the necessary number of hours to equalize batteries in a battery bank.

To start equalization, the inverter is switched either manually or automatically to the equalization mode and the generator is powered up, once again, either manually or automatically, depending on the type of inverter and generator. The inverter then controls the process. If the batteries are in a low state of charge when you want to equalize them, the battery charger will begin by bulk charging. Charging will then enter the absorption stage, but continue for an additional two to three hours at the same or a slightly higher level, to be sure the batteries have been equalized. Equalization time varies depending on the type of battery used in the system. Equalization drives lead sulfate off the plates, equalizes the voltage of all cells in the battery bank, and stirs the electrolyte. Once equalization is over, the generator either shuts off automatically or is manually shut down. ∎

energy in or out for a period of at least two hours (they need a little time without either for the voltage to settle). A 12-volt battery bank at rest is full when the voltage is about 12.6 volts. But the bank could have a voltage of 12.6 while the battery is under heavy charge, even if the battery is almost entirely discharged, and conversely, voltage of a fully charged battery may dip to 12.2 volts while carrying a large load such as a deep-well pump. For those who are interested in monitoring battery charge, manufacturers often provide charts that correlate state of charge with charge/discharge rate and voltage.

Individuals living with battery systems can gradually learn to interpret battery voltage, if they have no better tools at their disposal. A more precise way to monitor the state of batteries is with a digital amp-hour or watt-hour meter. One popular meter, the TriMetric, keeps track of the number of amp-hours of electricity moving in and out of a battery continuously. By doing so, and by it knowing the size of the battery bank in amp-hours, which is programmed into the meter, the meter can show how many amp-hours have been removed from the battery, but it can also show the battery's state of charge (SOC). (This meter uses an efficiency factor and formula to approximate the amount of energy lost to battery inefficiency.) To make your life easier, the TriMetric displays a "fuel gauge" reading that shows the state of charge as a percentage. Many laptop computers have similar state-of-battery-charge indicators for their batteries.

Homeowners can use battery information from meters to adjust daily activities. If batteries are approaching the 40 to 60% discharge mark, they may choose to hold off on activities that consume lots of electricity. Or, they may elect to run a backup generator to charge the batteries.

Another way to minimize deep discharging is to oversize a PV system — that is, to install extra modules or to install a hybrid system — for example, add a wind machine to supplement a PV array.

A backup generator can also be installed as a third source of electricity. Backup generators can be fired up when batteries are running low. Gen-sets can be wired directly to the inverter or charge controller so they start automatically when the battery voltage drops to a predetermined level. Bear in mind that not all inverters and not all generators operate automatically, so be sure your generator is compatible with this application and that your inverter contains the proper circuitry. For more on battery charging, see the accompanying sidebar "Understanding Battery Charging."

Watering and Cleaning Batteries

The lead plates in batteries are immersed in a 30% solution of sulfuric acid. As described earlier, sulfuric acid participates in reversible chemical reactions that store and release electrons. Because these reactions are reversible, you'd think that battery fluid levels would remain constant over the long haul. Unfortunately, that's not the case. Battery fluid levels decrease over time because water is lost when flooded batteries are charged. Why?

Electricity flowing into batteries splits water molecules in the electrolyte into their component parts — hydrogen and oxygen. This process is known as *electrolysis*. Hydrogen and oxygen produced during electrolysis are both gases. These gases escape through the vents in the battery caps in flooded lead-acid batteries (as they are meant to), lowering fluid levels. (Electrolysis is the source of the potentially explosive mixture of hydrogen and oxygen gas that makes battery room venting necessary.) Water can also evaporate through the vents, and a mist of sulfuric acid can escape through the vents during charging as well.

All of these potential sources of water loss add up over time and can run a battery dry. When the plates are exposed to air, they quickly begin to corrode. In our experience, a battery's life is pretty well over when this happens. Dan has tried all kinds of tricks to bring batteries that have fallen victim to low-electrolyte levels back to life, but to no avail. Even if only one cell in a battery goes bad, the entire battery is shot. Making matters worse, one bad battery renders a whole battery bank bad. If the battery bank is more than a year to a year and a half old, you can't simply replace a bad battery in the system with a new one. They all have to go. To prevent this potentially costly occurrence, monitor battery fluid levels regularly. Many experts recommend checking batteries on a monthly basis. Others recommend checking batteries every three months. Dan has found that a two-to-three-month check for battery fluid levels works well in his home and office.

Be careful not to get lulled into complacency after installing new batteries, however. The first year of a battery's existence is like a honeymoon. Peace and happiness prevail. Brand new batteries operate very smoothly for a year or so with very little water loss. As time passes, however, batteries become more demanding — a lot more demanding. They require more frequent inspection and much more care.

To check fluid levels, unscrew the caps and peer into each cell when the system is not charging. Use a flashlight, if necessary — never a flame from a cigarette lighter! Be sure the batteries are installed in a battery room or in a battery box so that they can be easily inspected. Avoid stacking batteries on shelves that preclude you from peering into the cells.

Battery acid should cover the plates at all times — at a bare minimum at least one-fourth of an inch above the plates. It's better to keep the battery more fully filled, but don't overfill it either. As a rule, it is best to fill batteries to just below the bottom of the fill well — the opening in the battery casing into which the battery cap is screwed.

When filling a battery be sure only to add distilled (or deionized) water.[8] Never use tap water. It may contain minerals or chemicals

Understanding Batteries

Sulfuric acid doesn't evaporate, but strong bubbling action during charging results in the production of a mist of acid that can escape through the vents in the battery caps. This is how acid gets on the top surface of the battery. Water may evaporate from a battery, but this accounts for very little water loss compared to electrolysis, the breakdown of water during charging, producing hydrogen and oxygen gases which escape through the battery vents.

that could contaminate the battery fluid and reduce a battery's life span.

Be sure not to overfill batteries. Overfilling a battery will result in battery acid bubbling out of the cells when batteries are charged. Water evaporating from the acid on the top of the battery leaves a white acidic deposit. It not only looks messy, it can conduct electricity, slowly draining the batteries. Battery acid also corrodes metal — electrical connections, battery terminals and battery cables. In addition, the loss of battery acid can result in a dilution of the electrolyte within the battery cells (the addition of distilled water to compensate for lost fluid will continually dilute the acid remaining in the cells).

Batteries should be fairly well charged before filling them with distilled water. Don't fill batteries, and then charge them with a backup generator. This could cause electrolyte to bubble out. (Battery fluid rises in the cell during charging due to the formation of bubbles on the plates and the warming of the electrolyte.) If fluid level is extremely low, however, add just enough distilled water to cover the plates *before* charging them.

Battery acid bubbling out of batteries needs to be cleaned promptly. Use distilled water and paper towels or a clean rag. Be sure to periodically clean up accumulations of sulfuric acid powder that form when acid mist escapes from the battery cap vents. Some sources recommend using a solution of baking soda (sodium bicarbonate) to neutralize battery acid that bubbles up onto the top of batteries. After carefully rinsing batteries with sodium bicarbonate, they say, batteries should be wiped down with distilled water again. We discourage people from cleaning batteries with baking soda because it could drip into the cells of a battery, neutralizing the acid and reducing battery capacity. It's been Dan's experience that distilled water is sufficient to clean the surface of batteries. When doing so, be sure to wear gloves and protective eyewear and a long-sleeved shirt — one you don't care about. If you get acid on your skin, wash it off immediately with soap and water.

Although we don't recommend using baking soda to clean batteries, it is not a bad idea to keep a few boxes of baking soda on hand, just in case there's an acid spill in the battery room. Acid spills are extremely rare but may occur if a battery case cracks. You should also install a dry chemical fire extinguisher for safety.

When filling batteries, be sure to take off watches, rings and other jewelry, especially loose-fitting jewelry. Metal jewelry will conduct electricity if it contacts both terminals of a battery. Such an event will leave your jewelry in a puddle of metal — along with some of your flesh. One 6-volt battery could produce over 3,000 amps if the positive and negative terminals of a battery are shorted. As if that's not enough, sparks could ignite hydrogen and oxygen gas in the vicinity. Heat produced by a short circuit could cause the battery case to explode ejecting hot battery acid in all directions.

Also, be careful with tools when working on batteries — for example, tightening cable connections. A metal tool that makes a connection between oppositely charged terminals on a battery becomes red hot and may be instantaneously welded in place. It can also ignite hydrogen gas, causing an explosion. Short-circuiting a battery will also very likely ruin the battery. To prevent these problems, be sure to wrap hand tools used for battery maintenance in electrical tape so that only one inch of metal is exposed on one end (that way it can't make an electrical connection). Or, buy insulated tools to prevent this from happening. Either way, it is best to have a dedicated set of tools for use only on the battery bank. Find a place for them next to the batteries. That way, you won't be tempted to use whatever (non-insulated) tools might be at hand at the moment you need them. Insulating all exposed battery terminals and cables with plastic is also a good practice to protect against accidental shorts.

You will also very likely need to clean the battery posts (terminals) every year or two. The fine corrosive mist containing sulfuric acid released when a battery charges corrodes the metal posts, battery cables and hardware.

To clean the posts, use a small wire brush and, if necessary, a spray-on battery cleaner. Cleaners can be purchased at a hardware or auto parts store. To reduce maintenance, coat battery posts with Vaseline or a battery protector/sealer, also available at hardware and auto supply stores. Both will protect the posts as well as the nuts that secure the battery cables on them.

Equalization

Recharging batteries and maintaining proper fluid levels ensures optimum function over the long haul. To get the most out of batteries, however, you also need to equalize them periodically.

Equalization is a carefully controlled and deliberate overcharge of batteries that removes lead sulfate from lead plates in batteries and extends battery life. Batteries can be equalized in several ways. In some systems, they can be equalized by the PV array, although this is difficult to do. A wind turbine in a hybrid system can also be used to equalize batteries. A long, strong storm, for example, can provide enough electricity to equalize a battery bank from a wind turbine. Equalization is most commonly carried out by running a backup generator. Periodic equalization is performed for three reasons: The first is to drive lead sulfate crystals off the lead plates, preventing the formation of larger crystals. They reduce battery capacity, are difficult to remove, and can permanently damage the lead plates.

Batteries must also be periodically equalized to stir up (de-stratify) the electrolyte. Sulfuric

Sulfate and Sulfation, What's Normal and What's Bad

During discharge in a lead-acid battery, lead sulfate, $PbSO_4$, is formed on the positive and negative plates. When first produced, lead sulfate forms a soft deposit on the plates. If the battery is promptly and fully recharged, lead sulfate is converted back into plate material (lead on the negative plates and lead dioxide on the positive plates) and sulfuric acid (electrolyte). However, if the battery is left discharged, even partially discharged, for some time, the lead sulfate slowly forms large crystals. These crystals are hard and are not easily converted back into plate material during a subsequent charge. An accumulation of these hard crystals is called *sulfation*.

Sulfation is bad for a couple of reasons. First, lead sulfate is an insulator. It reduces the chemical reactivity of the plates. Lead sulfate crystals therefore effectively reduce the surface area of the plates. In effect, batteries get smaller and their capacity to store energy is reduced. Second, hard crystals of lead sulfate eventually break off and fall to the bottom of the cell, reducing plate material. Not only is useful plate material lost forever, but the accumulation in the bottom of the cell eventually reaches the bottom of the plates. This prevents electrolyte from contacting the surface of the plates. At this stage, if not before, the battery is so degraded that it must be replaced.

acid tends to settle near the bottom of the cells in flooded lead-acid batteries. During equalization, however, hydrogen and oxygen gases released by the breakdown of water (electrolysis) create bubbles that cause the battery to "boil." (This phenomenon is called boiling because the rapid formation of bubbles resembles boiling in water; it is not caused by high temperatures.) These bubbles, in turn, mix the fluid so that the acid is de-stratified in each cell of each battery, ensuring better function.

Equalization is also performed to help bring all of the cells in a battery bank to the same voltage. That is, it equalizes voltage in the cells of a battery bank. That's important — indeed vital — to a battery bank because some cells may sulfate more than others. As a result, their voltage may be lower than others. A single low-voltage cell in one battery reduces the voltage of the entire string. In many ways, then, a battery bank is like a camel train. It travels at the speed of the slowest camel.

If equalization restores batteries, why won't a battery remain functional for an eternity? Although equalization removes lead sulfate crystals from plates, restoring their function, some lead dislodges — flakes off — during equalization, settling to the bottom of the batteries. As a result, batteries lose lead over time and never regain their full capacity.

Equalization is a simple process. In those systems with a gen-set for backup, the owner simply sets the inverter to the equalization mode and then cranks up the generator. The inverter or charge controller controls the process from that point onward. In wind/PV hybrid systems, the operator sets the wind generator charge controller to the equalize setting during a storm or period of high wind. The controller takes over from there.

During equalization, the charge current (number of amps fed into a battery bank) is kept relatively constant and high for a certain period, usually four to six hours. During this process, the battery voltage is allowed to rise to higher than normal levels. (Normally the charge controller shuts off or slows the flow of electricity to batteries once a certain voltage is reached.) Over time, the voltage of all of the cells of all the batteries in a battery bank rise to the same level (equalizing them). This process usually takes four to six hours. When the time is up, equalization is terminated automatically.

During equalization, voltage may rise quite high. For a 12-volt system, the voltage may rise as high as 15 to 16 volts. In a 24-volt system, it may rise to 30 to 32 volts. DC loads sensitive to high voltage should be disconnected during equalization. In AC-only systems, the inverter protects the system. (It regulates its output voltage to protect loads.)

How often batteries need to be equalized seems to depend on who you're asking. Some installers recommend equalization every three months. If your batteries are frequently deeply discharged, or if your battery bank consists of multiple series strings in parallel, you may want to equalize them more frequently, perhaps once a month. If batteries are kept topped off, that is, are rarely deeply discharged, they'll need less frequent equalization. For example, batteries that are kept pretty full and rarely or never discharged below 50% may only need to be equalized every six months.

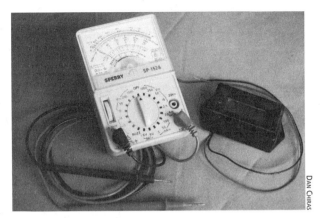

Fig. 7-11: *Voltmeters like these can be used to test the voltage of individual batteries in a battery bank. We prefer digital meters over analogue meters, as they're easier to read.*

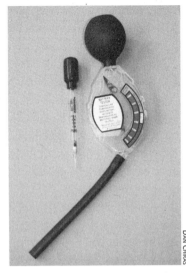

Fig. 7-12: *Hydrometers such as this one can be used to test the condition of batteries, one cell at a time. These devices measure the specific gravity of the electrolyte.*

The frequency of equalization also depends on your normal charge set points. Some homeowners run their battery banks "hot" — at a higher voltage setting than the standard. So the batteries have less need of equalization, but more watering.

Rather than second-guessing your battery's need for equalization, it is wise to check the voltage of each battery (using a voltmeter), every month or two. If you notice significant differences in battery voltage in a battery bank — that is, the voltage of one or two batteries is substantially lower than others — it is time to equalize the battery bank. Checking voltage requires a small voltmeter like the one shown in Figure 7-11. Dan uses a standard portable electrical meter like those used by electricians. A digital voltmeter is much easier to read than a dial meter.

Another way to test batteries is to measure the specific gravity of the battery acid using a hydrometer (Figure 7-12). Specific gravity is a scientific term for the density of a fluid. Density is related to the concentration of battery acid — the higher the sulfuric acid concentration, the higher the specific gravity. If significant differences in the specific gravity of the battery acid are detected in the cells of a battery bank, it is time to equalize.

Checking voltage of the batteries and the specific gravity of the cells may involve more work than you'd like and may not be necessary if you pay attention to weather and battery voltage or adhere to a periodic equalization regime. Dan rarely equalizes his batteries in the sunny summer months of Colorado, as his solar electric panels frequently produce copious amounts of electricity and his batteries are often full to overflowing with electricity. In the winter, however, he equalizes every two to three months. When batteries run low, he may run the generator for an hour or two to bring the charge up to prevent deep cycling. This is not equalization, just an attempt to prevent deep discharging.

Small wind system expert Mick Sagrillo recommends using a wind turbine to equalize batteries in off-grid systems, if possible. You'd be amazed at how well it works, as was solar electric expert Johnny Weiss of Solar Energy International in sunny Colorado. Johnny was astonished when he observed a newly installed wind turbine charge a battery bank in a PV/wind hybrid system. The wind turbine, installed during a workshop taught by Mick, was added because the client wanted to wean herself from a gasoline-powered gen-set. After living with the wind/PV hybrid system for a year, the owner sold the gen-set, as she no longer needed it. That's a properly sized PV/wind hybrid system! Unfortunately, this solution won't work for everyone. Conditions have to be right for this to work — that is, you need to have good solar *and* wind resources, and may need to be willing to curtail usage when both resources are scarce.

As a final note on the topic of equalization, be sure only to equalize flooded lead-acid

batteries. (Holes in the caps serve as vents that release hydrogen and oxygen gases.) A sealed battery, either the gel electrolyte or absorbed glass matt sealed battery, cannot be equalized! If you do, you'll ruin it. Also, don't equalize batteries fitted with Hydrocaps, discussed next. Hydrocaps must be removed prior to equalization or they will overheat and melt.

Reducing Battery Maintenance

Battery maintenance should take no more than 10 to 30 minutes a month. (It takes about 10 minutes to check the cells in a dozen batteries but may take a half hour to check them and add distilled water to each cell if battery fluid levels are low.)

If this sounds like too much work, you'll be delighted to know that there are ways to reduce battery maintenance. One way is to install sealed batteries, although they're best suited for grid-connected systems with battery backup.

Another way to reduce battery maintenance is to replace factory battery caps with Hydrocaps, shown in Figure 7-13. Hydrocaps capture much of the hydrogen and oxygen gases released by batteries when charging under normal operation. Hydrocaps contain a small chamber filled with tiny beads containing platinum catalyst. Hydrogen and oxygen escaping from the cells react in the presence of this catalyst to form water, which drips back into the batteries. Because of this, Hydrocaps reduce water losses by 50 to 90%, according to several sources, and thus greatly reduce battery filling.

Another option is the Water Miser cap. Water Miser caps are molded plastic "flip-top"

Fig. 7-13:
Hydrocaps shown here are special battery caps that convert hydrogen and oxygen released by batteries into water, which drips back into the cells of flooded lead acid batteries, reducing watering.

Joe Schwartz

A Note on Battery Cost

While battery replacement results in a significant cost, it's important to note that grid-connected systems also incur costs — monthly service charges. How do they compare? Suppose a battery bank for an off-grid home cost $2,700 and the homeowners are very good to their batteries and manage to get eight years of service from them. When time comes to replace the batteries, they suddenly have to shell out $2,700! Seems like quite a chunk of change for these folks getting their energy from the sun. However, the monthly service fee many grid-connected customers pay can also be quite substantial. Many utilities charge as much as $22 to $28 a month. If the fee were $28 a month, over an eight-year period, the cost would be equal to the cost of a battery bank. All an off-grid customer would have to do is deposit $28 a month into their own "customer account," so that when the battery bank needs replacement, the cash is at hand.

vent caps. Plastic pellets inside these caps capture most of the moisture and acid mist escaping from the battery fluid, reducing sulfuric acid fumes in the battery room and preventing terminal corrosion. They don't capture hydrogen and oxygen, however. While battery water loss is reduced, and less frequent watering is required, Water Miser caps only reduce water loss around 30%; they can, however, be left in place during equalization.

Of the two, we prefer Hydrocaps. They cost a bit more than Water Miser caps and must be removed before batteries are equalized, but they reduce water losses more than their competitor, which reduces the amount of filling you'll need to do.

Yet another — even better — way to reduce maintenance is to install a battery filling system, as shown in Figure 7-14. Battery filling systems consist of a series of plastic tubes connected to specially made battery caps (on each cell) fitted with float valve.

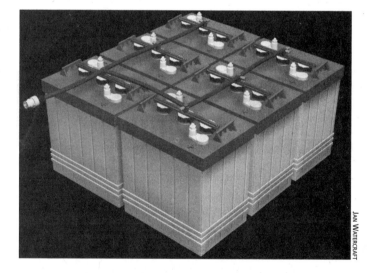

JAN WATERCRAFT

In automatic systems, the plastic tubing is connected to a central reservoir containing distilled water. Valves in the battery caps open when the electrolyte level in a cell drops. Distilled water flows by gravity from the central reservoir. When the cell is full, the float valve stops the flow of water. (Remember to

Fig. 7-14: *The Pro-Fill battery filling system greatly reduces the time required to fill batteries and make the chore much easier.*

Fig. 7-15: *Battery filler bottles work well for systems in which batteries are accessible.*

Although battery-filling systems work well, they're costly. Dan spent about $300 for a system that services his 12 batteries. Although a bit pricey, such systems quickly pay for themselves in reduced maintenance time and ease of operation. Dan finds that the convenience of quick battery watering overcomes the procrastination that leads to damaged batteries. Without them, it's easy to put off battery maintenance, which can end up costing you a fortune!

A simpler and cheaper alternative is a half-gallon battery filler bottle (Figure 7-15). It comes equipped with a spring-loaded spout (similar to the float valves in battery-filling systems). This handy device allows you to fill each cell precisely as the valve shuts off the flow of distilled water when the battery fluid level comes to within an inch of the top of the cell. You can order them from a local battery supplier or online from several suppliers for $10 to $20. To find a supplier, simply search for "battery filler bottles."

Charge Controllers

Now that you understand how batteries work and, perhaps even more important, how to take care of them, let's turn our attention to the charge controller, another device that helps us care for our batteries.

A charge controller is a key component of most battery-based PV systems. If you're installing an off-grid system or grid-connected system with battery backup, you'll need one. A charge controller performs several functions, the most important of which is preventing batteries from overcharging (Figure 7-16).[9]

keep the distilled water reservoir filled at all times!)

In manually operated systems, the plastic tubing on each string of batteries is connected to a plastic tube fitted with a hand-operated pump. One end is immersed in a container of distilled water, for example, a one-gallon jug of distilled water. A quick coupler allows the tubing to be connected to the pump and the distilled water supply. Water flows into each cell to the proper level. Pressure build-up tells the operator when all the cells have been filled to the proper level. The entire process takes less than five minutes.

Dan uses a manually operated Qwik-Fill battery watering system manufactured by Flow-Rite Controls in Grand Rapids, Michigan and sold online through Jan Watercraft Products. He's found that this system works extremely well and has turned battery maintenance from a chore to a pleasure.

How Does a Charge Controller Prevent Overcharging?

To prevent batteries from overcharging, a controller monitors battery voltage at all times. When the voltage reaches a certain pre-determined level, known as the *voltage regulation* (VR) *set point*, the controller either slows down or terminates the flow of electricity (the charging current) into the battery bank, depending on the design.

As shown in Figures 7-17a and b, charge controllers come in two basic designs, *shunt charge controllers*, and *series charge controllers*. As shown in Figure 7-17a and b, a shunt charge controller contains either an on/off switch or a resistor that short-circuits the array when the battery voltage reaches the voltage regulation

Fig. 7-16:
Charge controllers like this one (top left) from Apollo Solar regulate the flow of electricity to the batteries in off-grid and grid-connected systems with battery backup. Some charge controllers contain maximum power point tracking circuitry to optimize array output and other features as well, like digital meters that display data on volts, amps and electricity stored in battery banks.

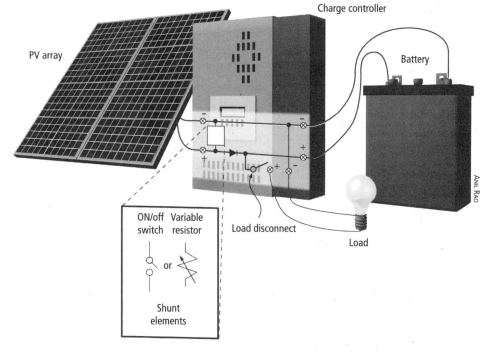

Fig. 7-17 a:
How charge controllers work. Shunt charge controllers short circuit the array through an on-off switch or a variable resister.

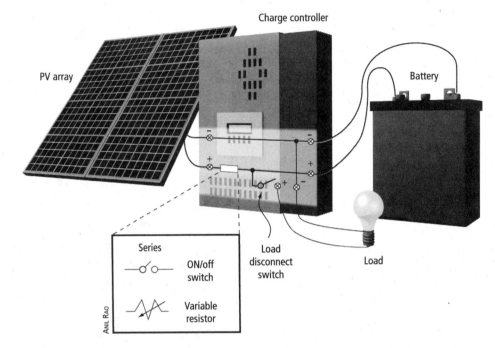

set point.[10] (The shunt is wired in parallel with the array.) When the shunt is activated, current flows through the shunt, and then back to the array.

A switch type controller is technically referred to as a shunt-interrupting charge controller, because the shunt completely interrupts the flow of current to the batteries. A resistor-type charge controller is technically referred to as a shunt-linear charge controller, because it contains a variable resistor that gradually reduces the flow of current. As the resistance in the shunt slowly declines, current flow through the resistor increases, and the flow of current to the batteries diminishes.

Both types of shunt controllers are said to close the circuit or short out the array, which protects the batteries from overcharging. How

long does the shunt operate? Current flows through the shunt until the battery voltage drops to a predetermined setting, known as the *array reconnect voltage set point*. In a shunt-interrupting charge controller, the switch opens entirely, allowing current to flow back to the batteries. In a shunt-linear charge controller, resistance increases, sending more current to the batteries.

The second type of charge controller is a series charge controller. As shown in Figure 7-17b, some series charge controllers contain an on/off switch wired in series. When open, it stops the flow of current to the battery bank. This type of controller is known as a *series-interrupting charge controller*.

The series element in a series charge controller can also be a variable resistor that

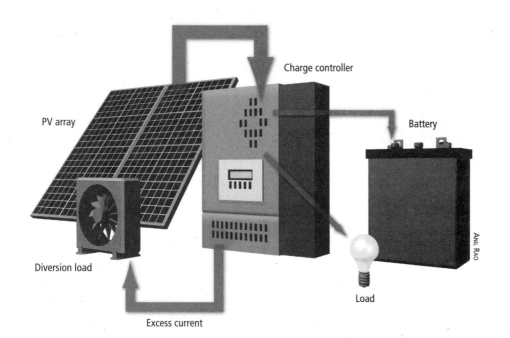

PV array

Diversion load

Excess current

Charge controller

Battery

Load

ANIL RAO

Fig. 7-18:
Diversionary charge controllers send surplus electricity to a dump load, either a resistive heater or fan or pumps, as explained in the text.

Preventing Reverse Current Flow

At night when the PV array is no longer producing electricity, current can flow from the batteries back through the array. To prevent this reverse current flow, charge controllers contain a diode in the circuit. Were it not prevented, reverse current flow could slowly discharge a battery bank. In most PV systems, battery discharge through the modules is fairly small, and power loss is therefore insignificant. However, reverse current flow is much more significant in larger PV systems. Fortunately, nearly all charge controllers deal with this potential problem automatically.

gradually increases resistance as battery voltage climbs toward the voltage regulation set point. This reduces the flow of current to the batteries.

While shunt and series charge controllers are the mainstay of the PV industry, battery-based wind and micro hydro systems rely on a third type of charge controller, a *diversion charge controller* shown in Figure 7-18. When batteries reach the voltage regulation set point, this type of controller sends surplus current to a diversion load.

A diversion load is an auxiliary load, that is, a load that's not critical to the function of the home or business. In wind-electric systems, it is usually a heating element. Heating elements may be placed inside water heaters or may be in wall-mounted resistive heaters that provide space heat. In PV systems, excess power is often available during the summer months during periods of high insolation. In these instances, the diversion load may consist of an irrigation pump or fan to help exhaust hot air from a building.

Diversion loads must be carefully sized according to the National Electrical Code, something an installer will be sure to do.

Why Is Overcharge Protection So Important?

Overcharge protection is important for flooded lead-acid batteries and sealed batteries. Without a charge controller, the current from a PV array flows into a battery in direct proportion to irradiance, the amount of sunlight striking it. Although there's nothing wrong with that, problems arise when the battery reaches full charge.

Without a charge controller, excessive amounts of current could flow into the battery, causing battery voltage to climb to extremely high levels. High voltage over an extended period causes severe gassing, water loss, and loss of electrolyte that can expose the lead plates. It can also result in internal heating and can cause the lead plates to corrode. This, in turn, will decrease the cell capacity of the battery and cause it to die prematurely.

Some overcharge is tolerated by a flooded lead-acid battery, so long as the fluid levels don't drop below the top of the lead plates and the electrolyte is replenished. But overcharge is especially harmful to sealed batteries. Overcharging them can cause water to escape through the pressure-relief valve — and there's no way for the water to be replaced. Overcharging can then damage the lead plates.

Overdischarge Protection

Charge controllers protect batteries from high voltage, but also often incorporate overdischarge protection, that is, circuitry that prevents the batteries from deep discharging. When the weather's cold, overdischarge protection also protects batteries from freezing. This feature is known as a *low voltage disconnect*.

Charge controllers prevent overdischarge by disconnecting loads — active circuits in a home or business. Figure 7-17 shows the disconnect switch.

Overdischarge protection is activated when a battery bank reaches a certain preset voltage or state of charge. This prevents the batteries from discharging any further. Overdischarge not only protects batteries, it can protect loads, some of which may not function properly, or may not function at all, at lower than normal voltages.

What Else do Charge Controllers Do?

Although the main purpose of a charge controller is to prevent overcharge and overdischarge of batteries, charge controllers often perform a number of additional functions. They may, for instance, control loads

— that is, switch loads on and off, depending on the time of day. This feature is commonly used to control PV-powered outdoor lighting, for example, to illuminate signs or parking lots. In these systems, the solar array essentially becomes a photo sensor. When the current or voltage from the array drops at the end of the day, the charge controller automatically switches on lights.

Some charge controllers may also be wired to outside sensors, for example, temperature or water-level sensors. Temperature sensors allow automatic control of cooling loads, for example, evaporative coolers. Water-level sensors are used to control irrigation pumps.

Finally, charge controllers may be designed to activate automated or user-activated equalizations. The charge controller in Dan's system, for instance, has a switch that can be turned to equalize the batteries from the PV array. During equalization, the charge controller raises the voltage regulation set point so the battery achieves a higher than normal charge. The charge controller maintains the higher set point for a pre-programmed period, allowing a very controlled overcharge, or battery equalization. While batteries can be equalized from a PV array, this process can be quite slow. It can take several days or even a week or two. The most rapid way to equalize batteries is through a gen-set.

Additional Considerations

Controllers are rated in amps; many controllers are rated at either 40 or 60 amps. The amp rating refers to the amount of current a controller can handle from an array. Thus,

arrays and controllers must be carefully matched. You wouldn't want to install a 40-amp controller in a PV system with an array that could produce 50 amps. Exceeding the amperage ratings on a controller could destroy the unit by overloading internal circuits.

The charge controller-rated current must be at least 125% of the maximum output of the PV array. This permits a controller to survive the occasional edge-of-cloud effect. This effect occurs when a cloud blocks sunlight, cooling down the array. When the cloud moves on, sunlight striking the array results in a spike in amperage, albeit a transitory increase. Also, if you suspect you might want to expand your solar electric system in the future, you may want to consider buying a larger charge controller.

Generators

Another key component of off-grid systems is the generator (Figure 7-19). Generators (also referred to as "gen-sets") are used to charge batteries during periods of low insolation. They are also used to equalize batteries and to provide power when extraordinary loads are used — for example, welders — that would exceed the output of the inverter. Finally, gen-sets may be used to provide backup power if the inverter or some other vital component breaks down. Although a battery-charging gen-set may not be required in hybrid systems with good solar and wind resources, most off-grid homes and businesses have one.

Generators are typically "hard wired" to the inverter, meaning that the operator doesn't need to brave the cold to start the generator every time he or she needs to fire the gen-set to charge the batteries. When the generator is started, either manually or automatically, the inverter "waits" about sixty seconds for the generator to warm up and stabilize, and then transfers all of the home loads over to the generator, (a transfer switch is located internally

Fig. 7-19: *Portable Gen-sets like these commonly run on gasoline.*

A Note on Generator Operation

When the generator is initially fired, often with the push of a button inside the home or business, the inverter "waits" a minute or so before connecting to it so that the generator has time to come up to speed and warm up a bit. Before turning the generator *off*, terminate charging, and let the generator run for a few minutes, so it has some idle run-time to cool off before being switched off completely.

within the inverter); it then performs a "smart," three-stage charge of the system's battery.

Gen-sets for homes and businesses are usually rather small, around 4,000 to 7,000 watts. Generators smaller than this are generally not adequate for battery charging.

What Are Your Options?

Generators can be powered by gasoline, diesel, propane or natural gas. By far the most common gen-sets used in off-grid systems are gasoline-powered. They're widely available and inexpensive. Gas-powered generators consist of a small gas engine that drives the generator. Like all generators, they produce AC electricity. It travels to the inverter (containing a battery charger) or to a dedicated battery charger. The battery charger contains a step-down transformer that reduces the 120- or 240-volt output of the generator to a voltage slightly higher than the nominal battery voltage, which could be 12-, 24- or 48-volts. The low-voltage AC electricity is then converted to DC electricity by a rectifier, also located in the battery charger. DC electricity is then fed into the batteries.

Gas-powered generators operate at 3,600 rpm and, as a result, tend to wear out pretty quickly. Although the lifespan depends on the amount of use, don't expect more than five years from a heavily used gas-powered gen-set. You may find yourself making an occasional costly repair from time to time as well.

Because they operate at such high rpms, gas-powered gen-sets are also rather noisy; however, Honda makes some models that are remarkably quiet (they contain excellent mufflers). If you have neighbors, you'll very likely need to build a sound-muting generator shed like Dan's to reduce noise levels, even if you do install a quiet model. And don't think about adding an additional muffler to a conventional gas-powered generator. If an engine is not designed for one, adding one could damage it.

Introducing the Inverter Generator

Some generators only produce full, rated output at 240 volts AC. At 120 volts AC, they only produce half their potential output. If the AC input of the inverter is 120 volts, it is recommended that a 240-volt generator be used in conjunction with step-down transformer . This allows the full output of the generator at 240 volts AC, which is transformed to 120 volts AC.

Full output can be important when we consider the way the battery charger in modern inverters work. In the last chapter, we discussed how the inverter transforms the lower battery voltage into higher "household" voltage, by having more windings on the output side of the transformer than the input or primary side of the transformer. If changing 12 volts into 120 volts, there would be a ten-to-one ratio on these windings. The transformer will work in either direction; if you put 12 volts into the transformer, 120 volts comes out, and if you put 120 volts in the other side, 12 volts comes out on the battery side. However, if we dig a little deeper into this process, we find ourselves wondering how we can charge a 12-volt battery bank to the 14 or 15 volts required to fully charge a battery with only 12 volts.

This happens because the 120 volts of our household current is the average voltage under the waveform and the voltage actually peaks at closer to 164 volts AC. This, in theory, makes it possible to charge a battery to 16.4 volts with this ten-to-one winding.

Unfortunately, the higher our battery voltage goes during charging, the "less" of the wave form the charger has to work with. What we then see happening, especially with undersized generators, is that the inverter starts out charging at a high current, with perhaps as much as 70 amps going into the battery, but as battery voltage rises during the charging process, current falls off. Higher quality and larger gen-sets will offer better peak voltages, while undersized and cheaper gen-sets tend to have the upper end (voltage) of the wave form "clipped," resulting in poor charging performance.

One type of generator that tends to offer good charging voltage is an "inverter generator," which is a DC generator that feeds an on-board inverter. Honda and others are making these quality gen-sets, and they are not only much quieter than conventional gen-sets, they will also provide more charging power, allowing for a smaller gen-set to produce superior charging power. Remember that the generator needs to be sized large enough to not only charge the batteries, but also to power the home loads at the same time. Gen-sets for homes and businesses are usually rather small, around 4,000 to 7,000 watts. Generators smaller than this are generally not adequate for battery charging.

If you're looking for a quieter, more efficient generator, you may want to consider one with a natural gas or propane engine. Large-sized units — around 10,000 watts or higher — operate at 1,800 rpm and are quieter than their less expensive gas-powered counterparts. Lower speed translates into longer lifespan and less noise. Natural gas and propane are also cleaner burning fuels than gasoline. Unlike gas-powered generators, natural gas and propane generators require no fuel handling by you. However, although they are quieter and produce less noise and pollution, natural gas and propane generators can be quite expensive. You could end up paying several times more for a natural gas or propane generator than for a comparable gasoline-powered unit.

Another efficient and reliable option to consider is a diesel generator. Diesel engines tend to be much more rugged than gas-powered engines. As a result, these heavy, long-lasting machines tend to operate without problems and for long periods. Less maintenance means lower overall operating costs and less hassle. Diesel generators are also more efficient than gas-powered generators. One big advantage of diesel engines is that they can be operated on biodiesel. Biodiesel is a fuel made from vegetable oil. It is typically mixed with petroleum diesel fuel in a ratio of 80% petroleum diesel to 20% biodiesel for cold-weather operation. In warmer weather, the mixture can be as high as 80% biodiesel and 20% conventional diesel.[11] (This is possible if you are making your own biodiesel; commercial blends are typically 20% biodiesel.)

Biodiesel burns cleaner than traditional diesel and is at least partly renewable because it's made from renewable resources, typically oils extracted from soybeans or canola.

Although diesel generators offer many advantages over gasline-powered generators, they cost more and, of course, you will have to fill the tank from time to time. They're also not as clean burning as natural gas or propane gen-sets.

Controlling Noise

William Kemp, author of *The Renewable Energy Handbook for Homeowners*, notes, "Aside from having to run a generator in the first place, the second most annoying feature of a gen-set is having to listen to it running." He goes on to say, "The best way to eliminate this problem is not to operate them at all." Although that's a great idea, it's difficult to achieve. In most cases, the best you can do is to minimize run time. Do this by selecting the sunniest location on your property for your PV array. Avoid shading. A tracking array can also help increase an array's output to keep batteries charged. You can also slightly oversize the array to produce more energy than you expect you'll need. Installing a wind turbine to provide additional power, even to equalize the batteries, can also help reduce generator run time. But, whatever you do, don't skimp on your generator run time to avoid hassle or eliminate noise. If you do, you could end up sending your battery bank to a premature death.

The next best alternative to not running a gen-set or reducing run time, Kemp goes on to say, "is to locate the unit a reasonable distance from the house and enclose it in a noise-reducing shed." Be sure not to operate a generator inside an attached garage or too near your home. Dangerous fumes can seep in.

To dampen noise, insulate the walls and ceilings. However, be sure to create an opening to remove exhaust and provide fresh air. Dan built a shed from leftover 2 x 6s and insulated the walls with a mixture of straw and clay. The ceiling was insulated with a mixture of wood chips and clay. He then applied earthen plaster to the inside and finished the exterior with a coat of lime-sand plaster. The floor of his shed was made from soil-cement, a mixture of subsoil from his property and 5% cement. He installed a louvered fan near the exhaust outlet of the gen-set. The exhaust fan is run off one of the 120-volt AC legs (outlets) on the generator. He created an air intake on the opposite side.

Kemp recommends building a shed with a floating floor — that is, a floor that does not contact the walls. This keeps engine vibration from radiating outside the building. "The walls should be packed solidly with rock wool, fiberglass or, best of all, cellulose insulation. The insulation should then be covered with plywood or other finishing material, further deadening sound."

Installing a gen-set away from your home may require long-distance feed cable run overhead or underground. The cable will need to be sized correctly to reduce line loss. The cable will need to be rated for burial or will need to be installed in conduit. Be sure not to install a gen-set close to a neighbor's house to avoid noise in your own home.

A generator with an automatic start switch will make your life a lot easier. This fairly inexpensive feature can be wired to the inverter (for truly automatic function) or to a manual switch inside the home. This feature will save you trips to the generator shed on the freezing cold winter days when the generator is most frequently called into active duty. Be sure to run wire for the automatic start circuitry when you run the power cable.

Yet another way to reduce noise is to install a quieter gen-set, for example, a gasoline-powered generator with a good muffler and a sound-absorbing cabinet, or a lower-rpm diesel, propane or natural gas generator. If there are no neighbors to bother, it may not be necessary to install a quiet generator in an insulated shed. However, be sure the generator is protected from weather.

Other Features to Look For

Gasoline-powered generators typically come with four 120-volt AC outlets into which extension cords can be plugged. They also often contain a 120-volt 20 or 30 amp outlet that is used to power battery chargers.

Dan's Coleman generator came with a highly desirable low-oil shutoff feature. It prevents the generator from starting if the oil level is low or shuts it off if oil levels fall below a critical level during operation. His older, but more expensive Onan generator simply has a low-oil indicator light.

Another feature to consider when shopping for a generator is a run-time meter. It will cost a bit more, but will help you keep track of run time so you can perform necessary maintenance.

Living with Batteries and Generators

Batteries work hard for those of us who live off grid. As you have seen, flooded lead-acid batteries need to be kept in a warm place, but not too warm. Flooded lead-acid batteries enclosures need to be vented, too, and kept clean. They also need to be periodically filled with distilled water. And, as if that's not enough, you will need to monitor their state of charge and either charge them periodically with a backup supply of power when they're being overworked or back off on electrical use. Either way, don't allow batteries to sit in a state of deep discharge for more than a few days.

Generators in off-grid systems need a bit of attention, too. If you install one in your system, you will need to periodically change oil and air filters. If you install a manually operated generator, you'll need to fire it up from time to time to raise the charge level on your batteries or to equalize the batteries. It is also a good idea to run a generator from time to time during long periods of inactivity, for example, over the summer when a generator is typically not used. Gasoline goes bad sit-

ting in a gas tank, too, so you may want to add a fuel stabilizer to the tank during such periods. If gasoline evaporates from the carburetor, expect a major repair bill. The residue forms a shellac-like material that really gums up the works. Finally, gasoline-powered generators can be difficult to start on cold winter days, so be sure to use the proper weight oil during the winter.

Gas-powered generators, while inexpensive, tend to require the most maintenance and have the shortest lifetime. Be prepared to haul your gas-powered gen-set in for an occasional repair. Dan's backup generator has been to the repair shop twice in 13 years for costly repairs — and he only runs it 10 to 20 hours a year!

In grid-connected systems with battery backup, you'll have much less to worry about. If you install sealed batteries, for example, you'll never need to check the fluid levels or fill batteries. Automatic controls keep the batteries fully charged.

Batteries may seem complicated and difficult to get along with, but if you understand the rules of the road, you can live peacefully with these gentle giants and get many years of faithful service. Break the rules and it's a sure thing you'll pay for your inattention and carelessness.

CHAPTER 8

MOUNTING A PV ARRAY
FOR MAXIMUM OUTPUT

For a solar electric system to operate at its full capacity, it must be installed correctly. First, the array must be mounted in a location that receives as much sunlight as possible throughout the year. As noted in Chapter 4, installers use a Solar Pathfinder or a similar device to select the sunniest — least shaded — locations for arrays. For best results, the array should be oriented so that it points directly to true south in the Northern Hemisphere or true north in the Southern Hemisphere. If the array is fixed, that is, if its tilt angle can't be adjusted, it should be mounted at an angle that provides the maximum output during the year. Or, the array can be installed so that it can be manually or automatically adjusted to optimize output.

Although the task of siting a solar array in a sunny location and orienting the array just right so that it generates the most electricity possible may seem pretty straightforward, it

can be quite challenging. For example, it is not always possible to find a location, especially on existing buildings, that provides full access to the Sun all year long. Shade trees, neighboring structures, or even portions of a building on which the array is mounted can shade it part of the time, reducing its output. While every effort should be made to install a PV array in a shade-free location, compromises are often necessary.

Siting an array may also be influenced by aesthetic considerations. Building departments, historic preservation districts, homeowners' associations and neighbors may influence the decision, putting pressure on you to place an array in a location that could lower its output. Dan had a client whose building department insisted that their array be mounted on a north sloping roof, so it wouldn't be so visible from the street that ran to the south of a new home they wanted to build in an historic district in

Chicago. (The client and Dan worked with the building department to help them understand why this location wouldn't work, but they eventually lost the battle and the client decided to sell the property and build elsewhere! Happily, Chicago is seeing a few more panels go up in non-historic neighborhoods.)

This chapter covers high-performance installation — that is, installation that will ensure a safe, productive PV array installation. It will introduce you to a number of installation options that provide some flexibility in meeting the twin goals of maximum output and aesthetics.

What Are Your Options?

Mounting options for the PV array are usually dictated by sunlight/shading issues at the site and by aesthetics. Generally speaking there are two options: roof mounted arrays or ground mounted arrays. Roof or building mounted arrays include "building-integrated," wherein the array is part of the roofing, and roof-racked arrays. Ground-sited arrays include those that are mounted on top of a steel mast or pole, and those mounted on racks attached to concrete piers. In this chapter, we'll discuss the various options within these two broad categories, beginning with pole-mounted arrays. We'll discuss each type and their pluses and minuses.

Pole Mounts —
Fixed and Tracking Arrays

A pole-mounted array is a PV array that's attached to a sturdy steel pole anchored securely in the ground, either in a hole made by an auger, as in the case of a smaller array, or in a concrete base, as is common in larger systems (Figure 8-1). The array is typically mounted on the top of the pole, although it can also be mounted on the side of the pole.

More on Pole-Mounted Arrays

Poles for pole-mounted arrays range from 2½-inch schedule 40 steel pipe for small arrays up to 8-inch schedule 40 or schedule 80 for larger arrays. (The term "schedule" refers to the thickness of the pipe wall and hence the strength of the pipe.) The pipe size (measured in inches) refers to the inside diameter of the pipe. Thus, a 6-inch schedule 40 pipe has an inside diameter of 6 inches. Its outside diameter (OD) will be about 6-5/8 inches. Rack manufacturers will specify the diameter and the schedule rating of the pipe needed to fit a specific rack. Because pipe is expensive to ship, installers purchase the pipe locally. Steel pipe is widely available from local steel or plumbing suppliers. Square steel tube is also used for larger arrays. Square tube won't turn in a concrete base when the winds blow on an array and is therefore better than round pipe.

To be sure that the array is firmly anchored in the ground, the steel-reinforced concrete base must be properly engineered. The concrete must also be poured a couple of weeks *before* the array is mounted on the pole so it cures properly.

Top-of-pole mounts usually offer greater flexibility with regard to mounting angle and enable operators to adjust the tilt angle of the array seasonally to accommodate changes in the Sun's elevation angle. Pole-mounted arrays also allow for tracking — automatic adjustment of the array so it follows the Sun as it "moves" across the sky.

Pole-mounted arrays are often fixed, that is, mounted so the azimuth and tilt angles of the array remain constant throughout the year. Installers choose the optimum year-round angles in such cases. As you may recall from Chapter 4, you can obtain data on optimum angle from the NASA Surface Meteorology and Solar Energy website.

To increase the output of your array, you may want to consider an adjustable pole-mounted array — also called a seasonally adjustable pole-mounted array. A manually adjusted array allows the owner to change the tilt angle of the array by season to increase the array's output. Although you can adjust an array four times a year — spring, summer, fall and winter — most people make the simple adjustments twice a year. The first adjustment is made on or around the spring equinox (the first day

Benefits of Adjusting Tilt Angle

Besides increasing output, tilting an array to take advantage of the low-angled winter Sun also allows the array to shed snow better than a flatter array would. Arrays at higher latitudes that are mounted at very steep angles may also receive energy from sunlight reflecting off snowy landscapes.

Fig. 8-1:
Fixed Pole-mounted Array. Each array is mounted on a pole anchored to the ground at the optimum angle for its location. Pole-mounted arrays provide some flexibility in placing an array on a lot.

ANTHONY POWELL

of spring). At that time, the array is adjusted at an angle equal to the site's latitude minus 15 degrees. Reducing the tilt angle positions the array so it captures more of the high-angled summer sun. The second adjustment is made on or around the autumnal equinox (the first day of fall). At that time, the array is tilted to an angle equal to the site's latitude plus 15 degrees.

This adjustment positions the array so that it more directly faces the low-angled winter Sun.

Seasonal adjustments of the tilt angle can increase the output of an array by 10% to nearly 40%, depending on the latitude of the site, location of the array on a piece of property, and shading. The adjustment is as simple as loosening a nut on a bolt on the back of the

Tilt Angle: Getting it Right

Most solar experts recommend adjustments twice a year: the first at or near the beginning of summer, the second at or near the beginning of winter. The rule of thumb is that during the summer the tilt angle should be set at latitude minus 15 degrees. If you live at 40° north latitude, for instance, the tilt angle should be about 25° in the summer. To maximize winter output, the tilt angle of the array should be set at the latitude plus 15 degrees. If your home or business is at 40° north latitude, then the tilt angle should be about 55° in the winter.

Although the rules of thumb work pretty well, we recommend that you check the NASA tables for slightly more precise recommendations. The NASA data is based on actual measurements and factors such as local cloud cover. As an example, Table 8-1, shows a site in south central Tennessee at 35° north latitude. According to the rule of thumb, the optimum year-round tilt angle at this location should be 35°; so the best wintertime tilt angle would be 50°, and the best summertime angle would be 20°. According to the NASA site, however, the optimum angle (found on the last row of the table) for best year-round performance for a fixed array is actually

30°. In the winter, the optimum tilt angle is 54 to 58° — slightly higher than the 50° general rule of thumb. In the summer, the optimum tilt angle is 2 to 13° — much lower than the 20° recommendation. This is due, in part, to the fact that there is more sunlight in the summer months. Adjusting the array so that it is "flatter," thereby capturing more of that higher summer sun, slightly increases output.

The performance difference between the two methods of determining tilt angle is small. For example, using the data from Table 8-1 for the site in Tennessee, if you were to adjust the tilt angle according NASA recommendations, you would get a 0.7% improvement over the rule-of-thumb tilt angle in the winter, and a 1.7% improvement in the summer. Even though the difference is small, over the long term a small increase results in thousands of extra kilowatt-hours of electricity.

As noted in Chapter 2, adjustments to accommodate the altitude and azimuth angles affect the angle of incidence of the incoming solar radiation. The more directly the array is facing the Sun, the lower the angle of incidence and the greater the amount of solar energy available to the array. ☞

array mount, then tilting the array up or down (Figure 8-2). A magnetic angle finder is used to get the tilt just right. Once the angle is correct, the nut is tightened. See the sidebar, "Tilt Angle: Getting it Right" for advice on tilt angle adjustment.

A third — and very alluring — option for a pole-mounted array is a tracker, a mechanism that moves the array so that it points directly at the Sun (or as close as possible) at all times. Two types of trackers are available: single-axis and dual-axis.

A *single-axis tracker* adjusts one angle, the azimuth angle, hence its name. Single-axis trackers rotate the array from east to west, following the Sun's azimuth angle from sunrise

As the angle of incidence increases, two factors come into play: energy density and energy transmission. A high angle of incidence decreases energy density, and, as the angle of incidence increases, the amount of energy transmitted through the glass decreases and the amount of energy reflected off the glass increases. Both energy density and energy transmission are not affected much by angles less than 20 or 30 degrees and only start to fall off sharply when the angle goes beyond 45 degrees. ■

Table 8.1
Monthly Averaged Radiation Incident
On An Equator-Pointed Tilted Surface (kWh/m²/day)

Lat 35 Lon -87	Jan	Feb	Mar	Apr	May	Jun	Jul	Aug	Sep	Oct	Nov	Dec	Annual Average
SSE HRZ	2.23	2.93	4.01	4.98	5.52	5.80	5.79	5.28	4.72	3.75	2.60	2.04	4.14
K	0.44	0.45	0.48	0.49	0.49	0.50	0.51	0.51	0.53	0.53	0.48	0.44	0.49
Diffuse	0.97	1.27	1.66	2.05	2.36	2.47	2.38	2.13	1.71	1.30	1.02	0.88	1.69
Direct	3.10	3.45	4.16	4.62	4.76	4.94	5.09	4.85	5.05	4.81	3.71	3.04	4.30
Tilt 0	2.20	2.83	3.96	4.95	5.49	5.76	5.76	5.25	4.65	3.72	2.55	2.03	4.10
Tilt 20	2.84	3.39	4.42	5.13	5.41	5.56	5.60	5.32	5.08	4.47	3.28	2.70	4.44
Tilt 35	3.15	3.61	4.51	4.97	5.05	5.11	5.18	5.08	5.10	4.76	3.63	3.04	4.44
Tilt 50	3.30	3.65	4.38	4.58	4.47	4.43	4.53	4.60	4.87	4.80	3.79	3.22	4.22
Tilt 90	2.84	2.86	3.04	2.69	2.33	2.21	2.28	2.55	3.15	3.69	3.21	2.84	2.81
OPT	3.31	3.65	4.51	5.13	5.52	5.76	5.76	5.35	5.12	4.81	3.79	3.24	4.67
OPT ANG	55.0	45.0	34.0	18.0	6.00	2.00	4.00	13.0	29.0	45.0	54.0	58.0	30.1

Fig. 8-2: *The tilt angle of seasonally adjustable pole-mounted array can be changed to accommodate the changing altitude angle of the Sun to increase electrical production.*

to sunset. In a single axis tracking array, the tilt angle is either fixed or manually adjusted two to four times a year to accommodate for seasonal differences in the altitude angle of the sun.

A tracker that moves the array to follow both the altitude angle and azimuth angle of the Sun is known as a *dual-axis tracker*. These trackers follow the Sun's azimuth angle from sunrise to sunset like single-axis trackers. They also adjust the tilt angle to follow the minute-by-minute and seasonal changes in the Sun's altitude angle. As you learned in Chapter 2, the altitude angle of the Sun increases from 0° at sunrise to a high point at solar noon, then decreases to 0° again at sunset. At noontime in the summer, the array may be nearly horizontal, which allows it to capture the high-angled summer Sun.

Dual-axis trackers are most useful at higher latitudes because of the long daylight periods in the summer. In the tropics, a single-axis tracker will perform as well as a dual-axis tracker.

Although trackers increase an array's annual output, the greatest impact on energy production occurs during the summer when days are longer and typically sunnier. Output improves less during the short, often-cloudy days of winter. For those installing grid-tied systems with annual net metering, a tracker's excess summer production helps offset winter's lower production. Trackers are less useful for off-grid systems, however, because summer surpluses are typically wasted. Once the batteries are full, the array's output has nowhere to go. (Arrays are either short-circuited or open circuited, as explained in the section on charge controllers in Chapter 7.)

Trackers track either actively or passively. Active systems use small electric motors to adjust the array angles. Photo sensors mounted on the array send signals to a controller that activates the motors, causing the array angles

Fig. 8-3: *This electronic eye sends signals to the controller which adjusts the tilt angle and azimuth angle of the array to track the Sun across the sky.*

a

b

Fig. 8-4: *(a) This motor sits atop the pole and adjusts the azimuth angle of the array to track the Sun on its east to west path across the sky. (b) A smaller electric motor adjusts the tilt angle of the array to accommodate the ever-changing altitude angle of the Sun.*

to change as the Sun's altitude and azimuth angles change during the day. The electric motors may be powered by either (1) DC electricity directly from the solar array, (2) DC electricity from the battery bank, or (3) AC electricity from the house or business (it must be converted to DC electricity to run the motor at the array).[1] Most installers choose the last option: powering the motors with AC electricity. AC electricity is converted to DC by an AC-to-DC converter, typically mounted on the pole.

Figure 8-3 shows a photo sensor on an array Dan helped install as part of a hands-on workshop. Figure 8-4a shows the electric motor that adjusts the azimuth angle and Figure 8-4b shows the motor that adjusts the tilt angle. Some motorized trackers are controlled by computer programs that adjust the array day by day and hour by hour, based on the known position of the Sun at a specific location.

Passive systems require no sensors or motors and are preferred by many installers. They typically track only the Sun's azimuth angle, from east to west. Passive trackers are equipped with tubes positioned on either side of the array. They are filled with a liquid refrigerant — usually Freon. When sunlight strikes the tube on the left, it heats the liquid, causing it to vaporize and expand. Expansion, in turn, forces some liquid into the tube on the right. This causes the weight to shift to the right, which causes the array to rotate, tracking the Sun. These systems are touted as being dependable as gravity and reliable as the Sun, although, as you shall soon see, there are some limitations to this type of system.

Both active and passive trackers increase array output, but active trackers are more precise. That is, they do a better job of tracking the Sun. How so? Passive trackers follow the Sun from sunrise to sunset. At the end of the day, however, they remain pointing west. As the morning Sun rises in the east, it heats the unshaded west-side canister. This forces liquid into the shaded east-side canister. As liquid moves to the east-side canister, the tracker rotates so that the array faces east. Unfortunately, it takes an hour or two for the passive tracker to rotate back to the east. An active array, however, returns to its east-facing position right after sunset, so it begins producing electricity an hour or two earlier than a passive array the next morning. As a result, motorized arrays generate more energy over a given period than arrays with passive trackers. Passive trackers can also be deflected by wind, so on windy days they may not be pointing directly at the sun.

"Trackers are available for almost all sizes of arrays," notes Ryan Mayfield, in an article

How Important is Tracker Accuracy?

At the beginning of this discussion, we said that PV modules generate the most electricity when directly facing the Sun. This, of course, is true, but how much output is lost if the tracker is a little off — if it doesn't point the array *directly* at the Sun? The answer is — not much. The relationship between tracking error and output is a simple mathematical function, the cosine function. (The tracking error is the angle between a line perpendicular to the face of the array and a line from the array to the Sun.) As the following table shows, small tracking errors have very little effect on PV output.

Tracking Error (Degrees)	Relative PV Output
0°	100%
5°	99.6%
10°	98.5%
15°	96.6%
20°	94.0%
25°	90.6%

in *Home Power* magazine (Issue 104). For optimum performance, they must receive dawn-to-dusk sun. "There's no point in buying a tracker if your site doesn't begin receiving sunlight until 10 in the morning, or loses it at 2 in the afternoon due to shading from hillsides, trees, or buildings," Mayfield adds. So, which should you buy — a single- or a dual-axis tracker?

Even though a dual-axis tracker may seem like it would produce a lot more electricity than a single-axis tracker, the benefit is actually only marginal. As shown in Figure 8-5, a single-axis tracker generally results in the greatest improvements in array output. Dual-axis trackers increase output over a single axis, but only very slightly.

Should you buy an active or a passive system? The answer to this question depends on where you live. As a general rule, active trackers are more widely applicable. They work well in both cold and hot climates and everything in between. Passive trackers on the other hand, do not perform as well in cold climates, such as those of Minnesota and Wisconsin. They tend to be sluggish and imprecise in winter weather. That's because they depend on solar heat to vaporize the Freon. If you live in a cold climate, then, you may want to consider an active tracker.

In warmer climates, passive trackers are attractive options because they tend to be a bit more reliable. Active trackers rely on controllers and other electronic components such as motors and sensors. The more parts, the more likely that something will go awry. Remember, the only moving parts in a fixed

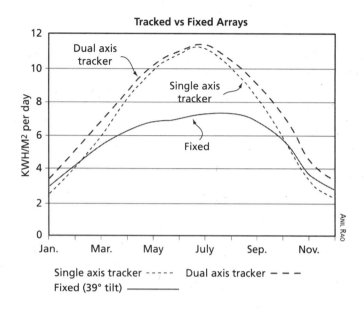

array PV system are electrons; when you add mechanical complexity to an otherwise simple design, you introduce wear, tear, maintenance and repairs. The more complex the system, then, the more problems you'll have. Photo sensors used to be the weak link in active trackers; they broke down frequently and required replacement. In recent years, however, manufacturers have improved their designs dramatically, resulting in much more reliable sensors.

Today, the weak link in active tracking arrays is the controller. They contain electronic circuitry that integrates information from the photo sensor to regulate the motors that adjust the tilt and azimuth angles of the array. Unfortunately, this circuitry is vulnerable to lightning. While manufacturers have improved lightning protection in active trackers, a

Fig. 8-5: *This graph compares the monthly output of a fixed array, a single-axis tracking array, and a dual-axis tracking array. A single-axis tracker dramatically increases the annual output compared to a fixed array. A dual-axis array results in very little additional gain.*

nearby or direct lightning strike can damage a controller, necessitating replacement.

According to some folks, the best answer to the question — should I install an active or passive system? — may be neither. Although it is true that trackers increase the output of a PV array, the key question to ask is how much does this additional output cost? And, is it worth it? Or, are there other, perhaps simpler and more reliable, ways to boost the electrical output of a PV array?

Proponents of the view that "no tracker is the best tracker" point out that, in most cases, you're better off investing in more PV modules to boost the output of a PV array than installing a tracking array that yields the same output. Both options, it turns out, cost the same initially. A larger, fixed array won't require a new controller when lightning strikes near your home.

If you're thinking about installing a tracker, run the math, or ask your installer to run it for you, to determine whether a tracker really makes sense. Be sure to take into account periodic controller replacement. And remember that shading at the site may make this a moot point. Trackers are only suitable for sites that have a horizon-to-horizon, open solar window.

Pros and Cons of Pole-Mounted Arrays

Pole-mounted arrays offer many advantages over other mounting systems. One of the biggest advantages of pole-mounted arrays is they can be positioned far away from objects that might shade an array — just so long as they are not too far from the balance of system.

According to PV expert Joe Schwartz, if the modules are wired to 48 volts nominal or higher, an array can be located a couple hundred feet from the inverter or batteries with minimal transmission loss and relatively small gauge wire.

Arrays mounted on poles can also be positioned precisely — that is, pointed south and tilted at just the right angle. Pole-mounts also help to maintain a cooler array than a roof mount. Most PV modules produce more electricity at cooler temperatures.

Precise positioning and cooler array temperatures increase the output of a PV array and increase the value of your investment in PVs. That is, you'll get more electricity out of the system over its lifetime, which lowers the cost per kilowatt-hour.

Pole-mounted arrays are also a lot safer to install than roof arrays. There's no need for potentially dangerous roof work. Pole-mounted arrays may be your only choice if a roof is unsuitable for some reason — for example, if the roof faces the wrong direction, is at an inappropriate pitch, is shaded, or is simply not strong enough to withstand wind loading (force of the wind that could rip an array from a roof). In addition, pole-mounted arrays do not require roof penetrations like roof mounted arrays. Driving lag screws into a roof to install a rack and cutting holes to run wires to the balance of system create potential leaks in a roof. Pole mounting an array, as opposed to a roof mount, also avoids the dismantling of an array when time comes to re-roof a house. Generally speaking, PV modules will long outlive most roofs, sometimes multiple re-roofings.

Pole mounts also permit easy access to solar modules, allowing an operator to clean a dusty array, if necessary, or gently brush snow off the modules to maximize output. (Cleaning is rarely required in most regions.) And, of course, pole mounting permits easy access for inspection and maintenance, although this is rarely necessary. PV arrays require very little, if any, maintenance over their long lifespan. The only maintenance Dan's had to perform on his roof-mounted array in 13 years was tightening a screw in a conduit bracket.

On the downside, pole mounted arrays are not usually suitable for small lots in cities and suburbs. Trees and homes can block the sun. Your own home, in fact, could block the sunlight striking a backyard pole-mounted array. Rural lots are typically better suited to pole mounted arrays. Even then, if a tracker is used, it needs to be positioned away from trees, barn and other buildings to ensure dawn-to-dusk tracking. As noted earlier, for optimum performance, a pole-mounted tracker array needs to be able to follow the Sun from sunrise to sunset, almost a full 270 degree path from northeast to northwest in June.[2]

Pole-mounted arrays are also more accessible to vandals. In addition, precautions need to be taken to prevent livestock, such as horses and cattle, from contacting the arrays. (This is usually solved by mounting the array on a taller pole or enclosing it within a fence.)

Pole-mounted arrays are also vulnerable to strong winds. An array becomes a large sail on windy days. Proper design and installation are critical to prevent wind damage.

Installing a Pole-Mounted Array

Pole-mounted arrays should be installed by a professional installer or a very experienced homeowner with electrical and PV experience. Installation will first require excavation of a hole to pour the concrete foundation. Exact specifications for the excavation and size of the pole are available through manufacturers of pole mounts. As an example, for a 120-square-foot array, Direct Power and Water, which manufactures pole mounts, recommends 6-inch schedule 40 pipe. For a pole that's 6 to 7 feet above ground, they recommend 4 to 6 feet of pipe underground, buried in the concrete foundation.

When installing a system yourself, be sure to check with the local building department. Some jurisdictions require homeowners to hire a licensed professional engineer to specify foundation details for local soil and wind conditions. If you're hiring a professional installer, they should secure permits. If you use round pipe, be sure to secure the pipe to the reinforcing steel in the foundation so it will not twist under wind loads.

You will also need to run wire from the house or business to the array. When soil conditions permit, the wire should be underground. Include separate conduits through the foundation for the DC output of the array and the power to the tracker. Once the pole is in place, you'll very likely need to mount a *combiner box* on it (Figure 8-6). It combines the output from each of the strings in a PV array into a single wire. The combiner box, in turn, is often wired to a *transition box*. The transition box allows smaller cable from the combiner box to connect to larger diameter wire that runs to the house. Larger diameter copper wire is used to reduce line loss over long distances. This wire enters a transition box inside the house, and is connected to the inverter in a grid-tied system or the controller in a battery-based system (with appropriate disconnects or breakers, of course, prior to reaching these destinations). Wiring the balance of system requires skill and knowledge. Wiring into a breaker box meter requires a licensed electrician in many locations.

Once the foundation has completely cured (don't rush it!), the mount is assembled on top of the pole. Most pole mounts are designed for specific modules, so be sure to specify the

KURT NELSON

Fig. 8-6 a and b: *The combiner box, shown in (a), combines the output of the various strings in a PV array for an off-grid system. A single wire runs from the combiner box to the house. Positive wires from the PV arrays connect to circuit breakers. Two large diameter wires run from the positive and negative bus bars (top and left, respectively) to the house. Combiner boxes for high-voltage grid-tied systems look slightly different as they typically contain fuses rather than circuit breakers.*

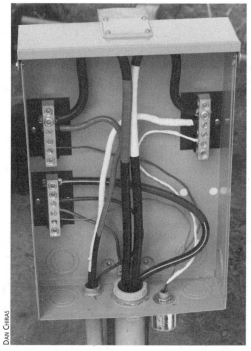

DAN CHIRAS

b: *In this installation for a grid-connected system in Minnesota, the array was located several hundred feet from the house, so the output of the combiner box was run to a nearby transition box, shown here. From here a single, large-diameter wire ran underground to the house. Large-diameter wire is used to reduce line loss.*

a

b

Fig. 8-7 a and b: *Pole-mounted array under construction. (a) Worker attaches the aluminum braces that will support the PV modules to a steel strongback (b).*

number and the exact modules you'll be using when ordering your pole mount hardware. Some installers recommend purchasing a larger rack than initially required so that more modules can be added at a later date if you decide to increase the size of the system. When installing a pole-mounted rack, or any rack for that matter, be sure to follow the instructions provided by the manufacturer.

If a tracker is used, you'll need to mount the motor unit on top of the pole, then install a cross piece, a heavy piece of steel that runs perpendicular to the pole. This piece, known as the *strongback* or *torque tube*, forms the spine of the mount. Cross braces are attached to the strongback (Figure 8-7) and the modules are then inserted one by one into the metal rack (Figure 8-8). Modules are typically wired in

Fig. 8-8: *Workers carefully slide PV modules into place on this dual-axis tracker at a hands-on workshop in Minnesota taught by Chris LaForge.*

series, creating strings that are then combined (wired in parallel) in the combiner box. This keeps the voltage the same but increases the amp output of the system.

Installing an array on a pole-mount can be challenging, so proceed carefully. Take a workshop or two and be sure to read Joe Schwartz's two articles on the subject in *Home Power* magazine (Issues 108 and 109). They're superb and very well illustrated.

Rack Mounts

Pole mounts are growing in popularity, but most PV systems are still mounted on racks. Racks, like the one shown in Figure 8-9, can be mounted on sloped or flat roofs. Racks are also used to mount PV arrays on the ground — secured, of course, to a solid foundation. Racks can even be mounted on the sides of buildings. There, the PV modules serve two functions:

they generate electricity and shade windows and walls from the intense summer sun, helping to passively cool homes and businesses.

Once made primarily from steel, racks are now manufactured from aluminum. These solid, durable racks secure the array to the roof, wall or ground, preventing the array from being carried away by winds. They may also provide a means of adjusting the tilt angle to optimize energy production.

Racks consist of three main components: (1) anchors or feet that attach the rack to a solid surface, a rooftop, wall or foundation in the case of a ground-mounted rack; (2) legs that allow the installer to set and, in the case of adjustable racks, to change the tilt angle seasonally; and

Fig. 8-9:
This aluminum rack made by UniRac can be mounted on roofs or on the ground. They may be fixed or seasonally adjusted. Note the mounting clips on the bottom rail.

DAN CHIRAS

(3) horizontal bars called rails to which the PV modules are attached (Figure 8-9).

Racks for solar arrays come in two basic varieties: fixed and adjustable. Fixed racks are typically set at the optimum angle, typically the latitude of a location.[3] If you live at 35° north latitude, for instance, the array should be set so the tilt angle is 35°. Be sure to check with the NASA website for the optimum tilt angle.

Adjustable racks, as their name makes clear, can be adjusted to maximize annual energy production by an array. Most arrays are adjusted seasonally, usually twice a year. Adjustability may be provided by telescoping back legs that allow the rack to be tilted up or down. Telescoping legs may have specific tilt-angle attachment points that permit a limited number of settings or may be infinitely adjustable. Some tilting legs come with predrilled holes or slots that correspond to different tilt angles.

Pros and Cons of Rack Mounting

According to Ryan Mayfield, one of the biggest advantages of rack mounting is that racks can be "designed to allow for a variety of tilt angles." As a result, "the PV array can be set at an optimal tilt angle based on the site's latitude or, if adjustable racks are chosen, repositioned to optimize energy output."

Racks are also quite versatile. A rack that can be mounted on a roof may also be mounted on the ground on concrete pilings or pressure-treated wooden posts driven into the ground. The same rack may also be suitable for an awning mount. Adding to their versatility,

many racks are designed to accept a variety of different modules — that is, different makes and sizes.

Another important advantage of racks is that they are relatively easy to install. Metal footings are first secured to the roof (Figure 8-10). Footings are usually either L-shaped brackets or post-type mounting feet that are lagged into the roof (Figure 8-10). To stream-line the installation and minimize time spent on the roof, racks are often assembled on the ground and then lifted onto the roof. The legs of the rack are then attached to the footings. Modules are secured to the rack either by mounting clips that are bolted into slots in the rack, as in Figure 8-9, or by bolts that connect the aluminum frame of the module directly to the rack from the backside.

Roof-mounted racks permit air to circulate around an array, keeping it cooler and

Fig. 8-10: *These L-footings are bolted into a 2 x 4 block attached to adjacent roof rafters. Silicon caulk is used to seal the hole and seal the base of the footing to prevent water from leaking into the roof.*

Fig. 8-11

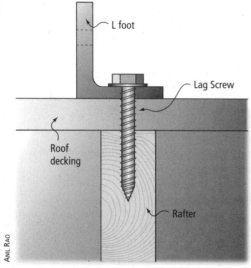

L foot

Lag Screw

Roof
decking

Rafter

ANIL RAO

increasing its output. Racks also allow ease of access to the back of the PV modules, making wiring easy. (Modules are typically wired in series and connected to a grounding wire. This work is performed on the back of the array.)

With roof-mounted racks, a PV array can be positioned well above obstructions, such as trees or neighbors' homes. This boosts the array's output. If the roof is shaded, ground-mounted racks may be a desirable alternative. In such cases, the array can be positioned in the sunniest location on a property. Ground-mounted arrays also make it easy to access an array for cleaning, snow removal, and inspection and maintenance.

On the negative side, racks are subject to the powerful forces of the wind. Rooftop, awning, and, to a lesser extent, ground-mounted, racks can be ripped from their moorings by strong winds — if not properly installed. (We'll discuss ways to prevent the

wind from having its way with your PV system shortly.)

Roof-mounted racks also result in penetrations in the roof. If not sealed properly, they can leak. Water leaking through the roof can dramatically reduce the R-value of ceiling insulation. (R-value is a measure of insulation's ability to resist heat flow.) Moisture can promote mold growth and can discolor and damage ceilings. If the leak is persistent, it can also damage structural members of the roof and ceiling, leading to serious problems.

Roof-mounted racks can also be difficult and costly to install in some instances — for example, on homes with steep or extremely high roofs. Expect to pay more, if you have one of those difficult-to-access and dangerous roofs.

Roof-mounted PV systems are also more difficult to access than ground-mounted or pole-mounted PV arrays to adjust the tilt angle, brush snow off the modules, or remove dust.

Another downside of racks, particularly roof-mounted racks, is that they can limit an array's exposure to the sun. For instance, if the roof of a home doesn't face true south, it may be difficult to mount a rack so that the array is oriented to the south. This is not a shortcoming of the rack, but of the roof orientation. Some loss of output will result. (Even so, notes Kurt Nelson, a solar installer who served as a technical advisor on this project, some sites are more open to the southeast or the southwest than to the south. Arrays pointed in those directions perform fairly well.)

One of the most significant disadvantages of roof-mounted racks is that they must be disassembled and removed when time comes to re-roof a house. This is costly and time-consuming work that should be carried out by a PV professional, not your local roofing contractor.

Removing an array to re-roof a house is also a huge annoyance. Asphalt shingles, found on the majority of homes in the United States live relatively short lives. This is especially true in areas that experience damaging hailstorms or areas exposed to intense sunlight or extreme temperatures. So, when contemplating a roof-mounted rack, carefully evaluate the condition of your roof or hire a trustworthy roofer to give you an honest opinion. If your roof needs replacement, or will need replacement soon, do so. To qualify for PV incentives in Oregon, homeowners proposing to mount an array on a roof that's ending its useful life must replace the shingles *before* they can receive incentives.

When building a new home or reroofing an existing home, we recommend installing a durable roofing material. A metal roof that should last for 50 years or more is an excellent choice, if it fits with the architectural style of your home. Many of the environmentally friendly roofs made from recycled plastic and rubber also outlast standard asphalt shingle roofs by decades. At the very least, chose high-quality (long-lasting) asphalt shingles. They're more expensive, but could last 40 years.

One type of metal roof, called a standing metal seam roof, has an added advantage of "providing" a ready-made rack. True standing seam roofs are somewhat costly, but very long lived and, utilize a clamp referred to as an S-5! clamp, the array can be mounted directly to the standing seams making for an installation that requires no roof penetrations.

Installing a Rack

Installing a rack on the ground or on a roof is pretty simple — if you've had some experience and know what you are doing. Workshop experience or the help of experienced friends can make the job a lot easier — and go a lot faster. Be sure to rope in to a secure point at all times so you or friends don't slide off the roof (Figure 8-12). (More on this shortly.) For best results, you should hire a qualified installer — one who has installed a lot of systems. They can do the job in a fraction of the time and the result may be much better.

Fig. 8-12: *To protect against falls, workers should clip their lanyards which are attached to safety harnesses and ropes securely attached to the building. Here, a worker secures climbing rope to the roof of a structure used to teach PV installation at the Midwest Renewable Energy Association.*

DAN CHIRAS

If you're going to do the job yourself, begin by carefully measuring your roof and laying out the array. Snap chalk lines to position the feet. Check your measurements two or three times *before* you start drilling holes in the roof. Be sure to work safely. Consider using scaffolding to access the roof, rather than a ladder. Scaffolding is a lot safer. Scaffolding will also make it easier to lift the rack and the modules onto the roof. Choose your shoes carefully. Sneakers with good tread may work well. Stiff work boots may not be as good on a roof. Stay off roofs when they're wet or storms are approaching. Remember that roofs heat up on sunny days and hot asphalt shingles become soft and may be damaged by walking or foot twisting.

A full harness and a safety rope or two are vital (Figure 8-13). A lanyard connected to your harness is used to click into the safety rope strung across the roof. And, of course, never work when you are tired or impaired. Never work alone, either. Besides providing an extra set of hands, a helper provides another set of eyes and another mind that could help you avoid mistakes and hazards.

To prevent an array from being blown off a rooftop on a blustery winter day, the feet must be securely anchored to the framing members of the roof — either the rafters or

Fig. 8-13: *Safety harnesses like the ones worn here by workers should be worn by all installers when working on roofs of homes and businesses. The workers are clipped into safety ropes secured to the roof.*

DAN CHIRAS

ceiling joists, depending on the type of roof. Lag screws can be driven into rafters and joists, but never into the roof decking (Figure 8-14). You'll need a good stud finder or access to the attic. When accessible, spend some time mapping the rafters or trusses. If it's difficult to lag directly into framing members of a roof, wooden blocks can be installed between the framing members, provided you can access the attic. Lag bolts can then be used to secure the rack to the roof (Figure 8-15). Blocking is not possible in existing homes with vaulted ceilings, as there's no access to the joists.

When installing lag bolts or lag screws, be sure to apply a long-lasting caulk, that is, a silicon caulk. Caulk prevents rain or melting snow from penetrating the roof and dripping on insulation. Get the best quality caulk possible. A few extra dollars spent on a high-quality caulk helps prevent leaks and could save you thousands in costly roof repair.

The tilt angle of the array has been discussed at some length. For many grid-tied systems it may be best to mount the array flush to the roof (same plane as roof but raised up slightly to allow for cooling). Aesthetically this is the most pleasing and concerns about wind loading are greatly reduced. If the angle is slightly less steep than ideal, what is lost in winter production may be mostly made up for with higher summer production.

To prevent a ground-mounted array from being blown into the next county, the legs must be firmly anchored into a properly engineered foundation — usually concrete piers or pilings set in the ground.

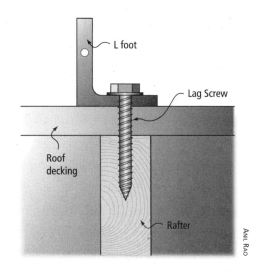

Fig. 8-14: *Properly sized lag screws are used to attach mounting brackets to the roofs of homes and offices.*

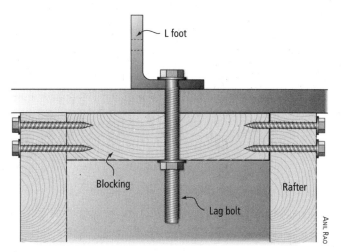

When assembling the rack, be sure to use stainless steel nuts and bolts. They'll last much longer than ordinary nuts and bolts. Modules are attached with clips or are bolted directly into the rack. The modules are wired in series using quick-connect adaptors, snap-in connectors that connect one module to the next. The positive and negative wires from

Fig. 8-15: *Blocks securely attached to adjacent rafters can be used to attach an array to a roof.*

Fig. 8-16: *Lugs (a) attached to both the aluminum frame of PV modules and the steel or aluminum rack secure copper grounding wire. A continuous ground is vital to protecting a system. (b) This drawing shows the layout of the continuous ground on a PV array.*

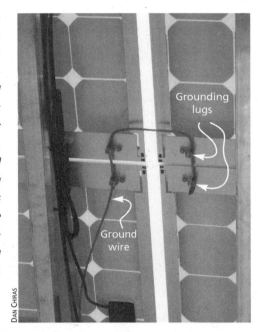

DAN CHIRAS

a

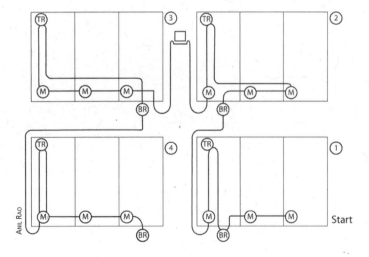

ANIL RAO

b

Key

BR = Bottom rail ground lug   M = Module ground lug
TR = Top rail ground lug

each string typically run to a combiner box, where they join the positive and negative leads from other strings (Figure 8-6a). Here the strings are wired in parallel. This wiring arrangement maintains the system voltage but increases the amp output of the array.

When installing a PV array, be sure that the array and the racks — and the rest of the system — are properly grounded. As Chris LaForge, veteran installer and instructor, notes in his writings on PV installation, "grounding is vital for safety and system performance." It protects users from shocks and also helps protect equipment from damage. Shock may occur if a wire carrying current comes in contact with a metal component of the PV system, for example, the frame of a module or the metal case of an inverter. If the system is not properly grounded, electricity could pass through the body of anyone who comes in contact with the shorted component.

When installing a PV array, you'll need to form one continuous ground. This is accomplished by installing lugs in the rack and each module in an array (Figure 8-16a). The grounding wire is run from lug to lug, then to the combiner box (Figure 8-16b).

To save both time and materials, Wiley Electronics and UniRack offer grounding clips (Figure 8-17). These simple devices clip in between adjacent modules on a rack, bonding the modules to the rails of the rack. This reduces the need to install grounding lugs on each module. Only one grounding lug is required for each rail.

Bare copper wire is used to ground an array. Installers typically use solid or stranded bare AWG 6 copper wire. This ground wire runs to the combiner box, then to an eight-foot ground rod, a.k.a. the grounding electrode, driven into the ground near the array. If the array is located some distance from the house, a ground wire is then run to the house where it is connected to yet another ground rod. The ground wire then enters the building where it is connected or bonded to the metal enclosures of the remaining components, for example the DC disconnects, inverter and controllers.

Grounding is both an art and a science, so be sure you know what you're doing. We strongly recommend that you consult with a qualified solar installer or hire a solar installer or electrician to work with you to ground your system properly.

One final note on installing an array on a rack. Be sure to do a neat job. It looks much better! Loose wires should be attached to the rack with black (UV-resistant) zip ties.

Standoff Mount

Your third option for mounting an array is known as a *standoff mount*. Shown in Figure 8-18, this special form of roof mount allows installers to mount PV arrays parallel to the roof surface to reduce their profile and visibility. Because the array is mounted parallel to the roof surface, the PV modules are mounted at a fixed tilt angle corresponding to the angle of the roof.

In standoff mounts, the array is raised slightly off the roof by four to six inches, sometimes more. (The more the better from a performance standpoint.) The space between

Fig. 8-17: *The WEEB Grounding Lug by Wiley Electronic's connects the aluminum frame of the PV module to the rack, reducing the wiring required to properly ground an array.*

Fig. 8-18: *These modules are mounted "flush" to the roof to reduce their profile. They're actually mounted several inches off the roof on a metal rack. This is known as a standoff mount.*

the roof and the array helps keep the PV modules cooler than a direct mount or an array attached directly to the roof (an approach that is rarely, if ever, seen except when laminate PVs are attached to standing seam metal roof).

"Standoff-mounted arrays are the most common, preferred, and least-expensive method for installing arrays as a retrofit or to existing rooftops," according to Jim Dunlop, author of *Photovoltaic Systems*, an excellent textbook on solar electricity for professionals. The arrays are mounted on rails typically attached to metal feet lagged into the roof framing members.

Pros and Cons of Standoff Mounts

Standoff mounts are popular among homeowners and businesses primarily because they minimize the profile of an array, making it appear as if it is part of the building. This option is not only aesthetically appealing, compared to pole or rooftop racked arrays, but it also reduces the wind's effect on arrays. It's far less likely that an extremely strong wind will dislodge a standoff-mounted array from its anchorage than a conventionally mounted array because the wind passes over the top of the array.

While aesthetically appealing, standoff-mounted arrays have their downsides. One of them is that the pitch (slope) of the roof determines the tilt angle. If the pitch of the roof does not correspond with the optimum tilt angle, the PV array's production could be seriously compromised. If the pitch is too shallow, for instance, wintertime production could be substantially lower than a homeowner could achieve by installing a properly angled rack or pole array. For off-grid systems, reduced output, especially in the winter, can

be devastating. To compensate, a homeowner would need to install a larger system — potentially, a much larger and more costly PV system. If, on the other hand, the pitch is too steep, the array's output in the summer could be compromised.

Another disadvantage of standoff mounts is that they frequently result in higher array temperatures than rack-mounted or pole-mounted arrays. This reduces the array's annual output.

A third problem, also encountered when installing roof racks, is that the PV array must be removed when a roof needs replacement.

And, even though the array is not mounted flush, it's difficult to access the backside of the modules. This may make it more difficult to wire a system during installation and certainly makes it more difficult to inspect an array once it is installed. (Modules can be removed to perform tests and inspect wiring, but access is limited.)

Installing a Standoff System

As in roof-mounted arrays, most standoff mounts are installed from the outside surface of the roof. Feet or standoffs are lagged into the rafters or joists. To mount some standoff systems, installers must access the underside of the roof. (This is impossible in vaulted ceilings of existing homes.) Once the feet are attached, the rails are secured. The PV modules are then placed on the rack and clipped in or bolted to the rails.

In buildings with standing seam metal roofs, however, the feet can be bolted to special clamps enthusiastically called S-5! clamps by the manufacturer. The clamps connect to the upright seam in standing-seam metal roofs. The rack is bolted to the clamp, creating a secure attachment without penetrating the roof. Four models of S-5! clamps are now available.

Building-integrated PVs

In Chapter 3, we discussed the many options for integrating PVs into buildings — for example, in roofs, walls, and glass. For most homeowners and businesses, there are three practical options: solar tiles, PV laminates on standing-seam metal roofs, and solarscapes. Let's take a brief look at each one.

Laminate PVs for Standing-Seam Metal Roof

If the roof of your home or business is fitted with a standing-seam metal roof, or if you're about to re-roof and would like to install this product, another option for turning your roof into an electric-generating plant is laminate PV (Fig 8-19). Developed by United Solar Ovonics, and the National Renewable Energy Laboratory, PV laminates are fabricated by applying multiple layers of amorphous silicon to a durable, but flexible metal backing. It is adhered to flat sections of standing-seam metal roofs, between the ridges (standing seams), and can be applied to new or existing roofs.

Installing PV Laminate

Uni-Solar's PV laminate (PVL) comes in rolls. (They're installed one at a time on sections of metal roofing. The ridge end of the roll is placed near the ridgeline. The paper backing is then removed from the first 16 inches of the roll, exposing the adhesive that secures the PVL to the roof. This section is then secured to the substrate, being careful to avoid creasing. The rest of the length is then rolled up toward the ridge. The backing is then removed a little at a time, as the PVL is unrolled down from the ridge to the eave.

When applying PVL, care should be taken to keep the laminate as straight as possible as it is unrolled. Uni-Solar's adhesive is pretty strong. You can't lift and reposition PVL if you mess up.

Once the laminate has been laid out and secured, installers run a hard-rubber roller over it, pressing it in place. (Installers use the same type of roller used to apply laminate to countertops.)

In new construction, PVL can be applied to the standing seam metal roof *before* the metal roofing is installed. This can be performed indoors on a temporary workbench. Doing so

Fig. 8-19:
PV laminate is applied directly to standing seam metal roof as shown here.

NATIONAL RENEWABLE ENERGY LABORATORY

gives you more control over the application and could result in a better installation. Be sure to follow the manufacturer's instructions.

Once the laminate is installed, the wire leads from the roof panels must be connected. Leads can then be run under the ridge cap if there is an attic or under eaves in buildings with vaulted ceilings. For ease of wiring, the manufacturer provides MC connectors, snap-in connectors like those found on standard PV modules.

Pros and Cons of PVL

PVL is easier and much quicker to install than conventional roof-mounted arrays. It blends with the building and from a distance may not be visible. Although PVL is flush mounted, thin-film products are less vulnerable to high temperatures than crystalline PVs — that is, their output doesn't decrease as much at high temperatures as monocrystalline and polycrystalline PVs. They're also not quite as sensitive to shading as standard PV modules. PVLs tend to be cost competitive with rack-mounted arrays.

On the downside: because PVLs on the market today are less efficient than crystalline PVs, you'll need twice as much roof space for the same system. Standing seam metal roof is, in our view, also not as attractive as other roofing materials. It's rarely used on homes, which limits the application of this product.

Solar Tiles

Individuals interested in BIPV (building-integrated photovoltaics) may also want to consider solar tiles, like those manufactured by Sharp, Atlantis Energy Systems, and Kyocera. Solar tiles incorporate crystalline silicon cells in a tile that is attached directly to the roof of homes and offices Figure 8-20. In SUNSLATES™, the single-crystal PV cells are mounted on a fiber cement slate backing. Kyocera's and Sharp's solar tiles are made from polycrystalline PV cells.

Like solar shingles, solar tiles do not need to cover the entire roof. You can, for instance, dedicate a portion of a south-facing roof to your array. The size of the system depends on the solar resource, roof pitch, shading and electrical demand. As in all solar electric systems, the ideal location is a south-facing roof; however, solar tiles can also be mounted on roofs that face southeast or southwest with only a modest loss (2 to 4%) in production, according to some manufacturers. Mounted on east- and west-facing roofs, the output could decline by 10 to 15%.

Like solar shingles, solar tiles overlap to shed water from the roof. They cannot be mounted on flat roofs. Roof pitch for SUNSLATES, for instance, must be at least 18 degrees (4/12 pitch). Be sure to check the pitch of your roof and manufacturers' recommendations before committing to a solar roof tile.

Solar tiles work well in a variety of climates, but perform best in sunnier locations. Be sure to check with the manufacturer for advice on the suitability of their product for your location. Also, be sure to check the compatibility of a solar tile with your roof. SUNSLATES, for instance, are "fully compatible with any other roof, be it tile, shake,

SUNSLATES

Fig. 8-20:
SUNSLATES from
Atlantis Energy
Systems are solar
tiles that replace
ordinary roof
shingles. The top of
this roof is fitted
with solar tiles that
generate electricity
to help meet
household
demands.

ATLANTIS ENERGY SYSTEMS

metal or composite," according to the manufacturer. The manufacturer may also have recommendations for the best combinations. Atlantis Energy Systems, for instance, states that their product goes best with gray concrete roof tiles.

Installing Solar Roof Tiles

Solar roof tiles are mounted on roofs over 30-pound roof felt. Each manufacturer provides directions. Kyocera's MyGen Meridian solar roof tiles are mounted on a metal rack that's flush mounted on the roof. SUNSLATES are mounted on a specially constructed wood framework — a series of vertical 2 x 2s secured to the rafters underneath (usually 2' on center). Horizontal 1 x 4s are nailed to 2 x 2s about 12 inches apart. Hooks are then nailed to the 1 x 4s and the slates are hung on them. Spacing the tiles off the roof creates a

space that allows air to flow beneath the slates, cooling them and helping to boost output. Shaded areas are fitted with blank tiles.

Solar roof tiles are grouped and wired together to form strings. SUNSLATES, for instance, are grouped in strings of 24. The positive and negative leads from each string, which are known as *home run cables*, are run into the attic to a combiner box. In most installations, the home run cables penetrate the roof at a single location — usually at the top center of the roof. A single cable runs to the inverter.

Pros and Cons of Solar Tiles

Solar tiles offer many of the advantages of other solar roofing products. Once again, though, the main benefit is aesthetics. They also have some of the same downsides, the most important of which is that they require a lot of connections. Many connections not only mean more work, it means that these arrays are more difficult to troubleshoot if something goes wrong.

Solarscapes

Another increasingly popular option designed to integrate solar electricity into new or existing homes and businesses is referred to as *solarscaping*. In this approach, PV modules are incorporated into the roofs of structures, such as pergolas, gazebos and carports. PV modules can be used to create awnings or shade structures for decks and hot tubs. The PV modules generate electricity while creating shade (Figure 8-21).

Solar shade structures are typically made from wood or steel and are designed and built so they blend in nicely with the existing architecture and landscape. These custom-made structures are erected in sunny locations where shade is desired or required.

Many installers use bifacial PV modules like those made by Sanyo. Bifacial modules were discussed in Chapter 3. This module was developed in the mid-1970s for use in spacecraft. Bifacial modules contain PV cells encapsulated — both front and back — by glass. This allows the module to harvest solar energy from the front side and the backside. Sunlight reflecting either off the ground or off light-colored materials, such as wood, metal roofing, concrete, gravel, snow and water, strike the backside of the PV cells, generating additional electricity. The glass-on-glass construction of bifacial modules allows some light to filter through the array, creating "a soft light-and-shadow pattern on the surfaces

Fig. 8-21:
PV modules can be used to create roofs of shade structures, a technique known as solarscaping. This photograph shows a portion of a car port made from PV modules by Lumos Solar.

LUMOS SOLAR

beneath the array," according to writer, Topher Donahue in an article on solarscaping in *Home Power* (Issue 122).

Pros and Cons of Solarscaping

Solarscaping is suited for new and existing homes and businesses. Like other types of BIPV, this approach helps minimize aesthetic concerns of conventional PV mounting systems, specifically racks and pole mounts. It therefore helps more people find a suitable way to incorporate PVs into their lives. "The more options we can provide our customers, the greater chance we can meet their needs," notes Scott Franklin, president of Lighthousesolar, a company headquartered in Boulder, Colorado. They manufacture the structures in their workshop, and then truck them to the installation site where they are assembled in one to two days, reducing time spent on a customer's property.

Solarscaping also offers the flexibility of pole-mounted and ground-mounted arrays — that is, it allows placement of arrays in sunny locations on sun-challenged properties. In addition, this approach offers ease of access to the array and ease of installation compared to roof-mounted arrays. And, by avoiding installation on the roof, you won't have to worry about removing the array when time comes to replace shingles. Like pole-mounted arrays, this approach results in a cooler and more energy-efficient array. Homeowners and businesses can even buy prefabricated solarscape kits from the Colorado-based company Lumos Solar (lumossolar.com).

Conclusion

You have many options when it comes to mounting a PV array. When considering your options, remember that one of your main goals is to produce as much electricity as possible. The accompanying sidebar, "Getting the Most from Your PV System" summarizes the many ways you can achieve this goal. Optimizing a PV array not only optimizes the value of your investment, it helps create a more sustainable supply of energy. But don't forget aesthetics and curb appeal. Although you may like the looks of a PV array mounted on a rack on a roof, neighbors might not be too keen. Future buyers may not like the look of it either.

Also, keep in mind the ease of installation. The more difficult and risky the installation, the more costly it will be. Although PV arrays require very little maintenance, access is especially important if you live in a dusty or snowy area. Arrays may need to be dusted off or have snow removed from time to time. Also, don't forget about winds and protecting an array from vandals and thieves.

Finally, solar electric arrays must be properly installed and must comply with all requirements of the local building department. The electrical portion of the installation must comply with the National Electrical Code and any local codes. If possible, you may want to consider making your array perform double duty, generating electricity and providing shade.

Getting the Most from Your PV System

1. Optimize electrical output by locating your PV array in the sunniest (shade-free) location on your property, orienting the array to true south, setting modules at the optimum tilt angle or installing on an adjustable rack or tracker. Ensure access to the Sun from at least 9 am to 3 pm.

2. Install a Maximum Power Point Tracking controller (for battery-based systems) or inverter for grid-connected systems to ensure maximum array output.

3. Keep modules as cool as possible by mounting them on a rack or pole, ensuring air circulation around the modules. Avoid flush-mounts (standoffs), if possible. In hot climates, install high-temperature modules, that is, modules that perform better under higher temperatures.

4. Install high-efficiency modules if roof space or rack space is limited.

5. Install modules with bypass diodes to reduce losses due to inadvertent shading.

6. When installing multiple rows, be sure to provide a sufficient amount of distance between the rows to avoid shading — that is, one row shading the row behind it.

7. Select modules with the lowest rated power tolerance. Power tolerance is expressed as a percentage, which indicates the percent by which a PV module will overperform (produce more power) or underperform (produce less power) the nameplate rating. Look for modules with a small negative or a positive-only power tolerance.

8. Install an efficient inverter. Use the weighted efficiency as your guide rather than maximum efficiency. Weighted efficiency is a measure of the efficiency under typical operating conditions.

9. Keep your inverter cool. If installed outside, be sure the inverter is shaded at all times. If installed inside, put it in a cool location. Be sure that air can circulate around the inverter.

10. Decrease line losses by installing a high-voltage array.

11. Reduce the length of wire runs whenever possible. Use larger conductors to reduce resistance losses.

12. Use quality (low-resistance) connectors and equipment. Be sure all connections are tight to reduce resistance and conduction losses.

13. Keep your modules clean and snow free. If you live in a dusty environment with very little rain, periodically dust off your modules. Remove snow from modules in the winter, but brush it off gently. Don't use any sharp tools to remove snow from a PV module. Mount the array so it is easily accessible for cleaning and/or snow removal.

14. For battery-based systems, maintain batteries properly. Periodically equalize batteries and fill flooded lead-acid batteries with distilled water. Recharge promptly after deep discharges and install batteries in a warm location so they stay at 75 to 80°F, if possible.

CHAPTER 9

FINAL CONSIDERATIONS:
PERMITS, COVENANTS, UTILITY INTERCONNECTION, AND BUYING A SYSTEM

To those who are enamored by the idea of generating their own electricity from the Sun, there are few things in the world more exciting than turning a PV system on for the first time and watching the meter show that electricity generated by the Sun is flowing into your home or business. It's an even greater thrill to see the utility meter running backward as surplus flows onto the electrical grid.

Although you may encounter a few obstacles along the way, in most instances, the path from conception to the completed installation is fairly straightforward. Table 9-1 summarizes the steps you or your installer must take in the order in which they must be completed. As you can see, we've already discussed steps 1 and 2 and details of others. In this chapter, we'll explore the remaining steps. We will begin by discussing permits, then look at restrictive covenants imposed by some homeowners' and neighborhood associations. We

will then discuss interconnection agreements required for grid-connected PV systems and insurance. We'll end with some advice on buying and installing a PV system.

Permitting a PV System

After you or your installer have designed your PV system, you'll need to contact your local government to determine if they require a permit to install the system. In most cases, a permit will be required. You'll most likely be directed to the local building department. They will provide a permit application and guide you through the process.

Building departments are typically the *authority having jurisdiction* (AHJ) over all aspects of construction in cities, towns and counties. They are granted authority and legal power to administer, interpret and enforce the building codes of local governments. Building codes are a detailed set of regulations that

Table 9.1
Steps to Implement a PV Energy Project

1. Determine your home or business' electrical consumption and consider making efficiency improvements to reduce the PV system size and cost.
2. Assess the solar resource.
3. Size and design the system.
4. Check homeowner association regulations; file necessary paperwork for permission to install the system, if required.
5. Apply for special incentives that may be available from the utility or your state or local government.
6. Check on building permit requirements; file a permit application.
7. Check on insurance coverage.
8. Contact the local utility and negotiate utility interconnection agreement (for grid-connected systems).
9. Obtain permit.
10. Order modules, rack and balance of system.
11. Install system.
12. Commission — require installer to verify performance of the system.
13. Sign interconnection agreement.

apply to how buildings are constructed, modified and repaired. Building codes set the rules by which the various trades must operate. They stipulate equipment and materials that can be used and how they must be installed. For example, they stipulate the way PV systems must be wired and the use of safety measures such as disconnects. Local trade licensing codes (separate from building codes) may also stipulate who can perform certain tasks, for example, licensed electricians or plumbers. Building codes also stipulate setbacks which limit how close a PV system may be installed to a public right of way or neighbors' property lines.

Building departments issue permits for structural and electrical work separately and

conduct inspections at certain stages of the work. When a project is completed and passes all required inspections, the local AHJ grants a certificate of occupancy, in the case of a building, or a certificate of approval in the case of a PV project.

Although AHJs can establish their own regulations for construction and remodeling, the process is expensive, time-consuming, and extremely costly. (In the United States, only Chicago and New York City have developed their own building codes.) Because of the time, labor and expense required to create a comprehensive building code, most AHJs have adopted model building codes, created by independent organizations. In the United States, for instance, most jurisdictions have adopted the *National Electrical Code* (NEC) to govern all electrical work on homes and businesses, including renewable energy systems. The National Electrical Code was developed by the National Fire Protection Association's Committee on the National Electrical Code. This group consists of 19 code-making panels and a technical correlating committee. Article 690 of the NEC applies specifically to the installation of photovoltaic systems.

Although states and cities typically adopt a model code, they are given the authority to modify the model code to meet local or regional needs. AHJs may waive certain requirements or permit alternative measures or equipment that ensures the same safety standards. Even so, AHJs rely almost entirely on the stipulations set forth in the NEC.

Building code requirements not only dictate how buildings are built, they also profoundly influence the manufacturers of electrical equipment, including PV systems. Manufacturers of all components of PV systems — from PV modules to inverters and DC disconnects — comply with code requirements — so the equipment they produce meets standards of the National Electric Code. (If it doesn't, it can't legally be installed in areas where building codes are enforced.) To meet NEC requirements, equipment must be approved by an independent testing laboratory, the largest of which is Underwriter's Laboratories (UL). It tests equipment and, if the equipment passes muster, then lists or certifies the product as meeting a certain standard. The UL listing is included on the nameplate of the product. UL listed products

may comply with US or Canadian safety standards, or both.

While the UL subjects PV equipment — including modules, inverters and charge controllers — to rigorous testing, they don't test every piece of equipment that rolls off the manufacturer's assembly line. Rather, they test representative samples and then routinely inspect production facilities for ongoing compliance with UL standards.

AHJs require the use of approved equipment in PV systems; plan examiners review permit applications, and inspectors verify that the proper equipment was used and that the work is done in compliance with the electrical code. Inspectors are licensed and certified by either the state or local government and have a thorough knowledge of building codes and construction, in the case of building inspectors, or electrical wiring, in the case of electrical inspectors.

Although there's a lot of grumbling about building codes and permits, they serve an important purpose. Building codes ensure that all building projects, including the installation of PV systems, are safe for the immediate occupants of a home or business and as well as for future residents of a home or employees of a business. The NEC, for example, protects against shocks and fires caused by electricity. If you install a PV system that is not in accordance with the local electrical code and it causes a fire that destroys your house, your insurance company may deny your claim. The NEC provisions that apply to PV systems also help ensure the safety of utility company employees, notably line workers, in the case of utility-connected systems. In a sense, then, building codes are society's way of ensuring the safety and well being of the present and future generations.

Although the requirements of local building codes can be daunting to the uninitiated, professional installers should be intimately familiar with them.

Securing a Permit

To start the process of ensuring that your PV installation will comply with local building code, you or your installer must submit an application for a permit. Permits for PV systems encompass electrical wiring and all the electronic components of a system you'll be installing, such as inverters, charge controllers, and safety disconnects. Your application may also be required to describe how the mounting of the array meets appropriate codes — for example, that an existing roof can support a roof-mounted array and that it is securely attached to the roof so it won't blow away in the wind.

To determine if a permit is required, even for off-grid systems, give your local building department a call. If a permit is required, they'll outline the procedure, indicate the cost, and provide the appropriate forms. They will also indicate all the supporting material you'll need to provide as part of your application.

Applications generally require a site map, drawn to scale, that indicates property lines, streets, and the proposed location of the array. This simple drawing should also indicate the location of other components of the system, for example, the inverter, combiner

box, disconnects and electric meter (in grid-connected systems). A professional installer will prepare a site map for you. It's part of his or her service.

Building permit applications for PV systems usually require a simple one-line drawing of the electrical system or, less commonly, a more complex three-line electrical diagram. Electrical schematics such as these indicate all of the components of PV systems and the connection with a home's or business' electrical system. You'll also need to submit specifications — size and ratings — of all the system components either on the system diagram or on separate specification sheets. This includes wire size, overcurrent protection devices (fuses or circuit breakers), system disconnects, and grounding equipment. You'll also need to include specifications for PV modules and inverters in grid-connected systems and charge controllers and batteries in battery-based systems. The building department will use this information to determine if the conductors, overcurrent protection devices, and disconnects are sized correctly.

Even if your local building department does not require an electrical permit, which is rare, the local utility will very likely require an electrical drawing and specifications— if you are connecting to the grid. They will examine the diagrams and specification sheets to be sure they comply with the NEC.

You will also very likely need to provide a description and/or drawings of the array-mounting design and the materials when applying for a permit through the local building department. If you are mounting an array on a roof, you'll need to include information on the age of the roof, type of shingles, the pitch, and the size and spacing of the rafters. You'll need to provide information on the array, too, including the weight and the method of securing it to the roof. You'll need to provide details on waterproofing the attachment as well. You may need to hire a professional engineer to review and stamp your plans. An engineer's stamp ensures that the roof can support the array and that it will remain intact under wind conditions in your area. If you're mounting an array on a pole mount, you will very likely need to provide a description and drawings of the foundation and pole mount to ensure its structural stability.

If you are hiring a professional installer, don't sweat it. The installer will file the permit and take care of these details. If you're installing a system yourself, however, you'll have to be sure that your plans comply with the local code and provide all the information required by the building department. If that sounds like too much, you may want to consider buying a packaged kit, which includes all of the components you'll need, including diagrams of the system and specifications of the components. Kits make it easier on the do-it-yourselfer.

After your application is submitted, a plan examiner in the building department will review the application and accompanying materials, usually within a few days or weeks, although the process can take several months, depending on the workload and staffing of the AHJ. If everything is in order, you'll receive a permit, an official approval for you to commence construction. (Never order

equipment or start work until you've received your permit!)

If your permit application doesn't meet the building code in some way, the plan inspector will mark up the schematics — it's called *redlining* — or provide written notice indicating the problem or problems and required changes. Once the problems have been resolved, you can resubmit the permit application.

Fees for permits for residential PV systems usually run from $50 to $1,500, depending on the jurisdiction. Some jurisdictions charge a flat fee; others charge a fee based on the size of the system.

Permits must be posted on the site, usually in a window. Permits often include an inspection schedule. It indicates which parts of the project are subject to inspection, what is covered in each inspection, and the sequence in which inspections must be carried out.

Expect an electrical inspector to visit your site at least once to check wiring and warning labels. (Warning labels are required on disconnects and other components of PV systems). The inspector will check out wire sizes, connections, overcurrent protection and grounding.

All equipment must be readily accessible to inspectors and must be installed according to code with proper clearances to ensure room to work on the equipment. This is known as *working space*. You'll also need to be sure that no other wiring, plumbing or ductwork is within a certain distance from your installed equipment.

A structural inspector may also visit to check out the array mount. You or the installer must call the AHJ to schedule all inspections, usually 24 to 48 hours in advance.

If you fail an inspection, you'll need to fix the problem and arrange a follow-up inspection. You may have to pay for follow-up inspections, too, although the cost may be included in the initial permit fee.

When applying for permits yourself, if your AHJ allows property owners to install their own systems, be sure you know what you are doing or hire a professional installer who does! Be courteous and respectful in all your dealings with the building department. If you are feeling irritated at having to file and pay for a permit, leave your angst at home. The same goes for dealing with inspectors. Inspectors have a tough job. They deal with difficult builders day in and day out and often show up in a sour mood. Be polite and respectful, even friendly — even if they fail your installation. Ask for explanations on ways to correct mistakes. They're usually happy to help out.

When applying for permits or dealing with inspectors, however, expect to be treated by building department officials in a reasonable and timely manner. A pugnacious attitude and list of demands are not well received. "From the official's perspective, there is no bigger turnoff than having to deal with an arrogant know-it-all," notes Mick Sagrillo, a veteran wind system installer. "On the other hand, don't be intimidated by resistance when dealing with building code officials."

While we are on the subject, avoid the tendency to bypass the law — that is, to install a PV system without a permit. Always obtain a permit when required. Do not,

under any circumstances, try to sidestep this requirement. The consequences are too great. Municipal governments have the authority to force homeowners to remove expensive unpermitted structures, even entire homes. Be sure to negotiate an interconnection agreement with the local utility, too. Mick knows people who installed grid-connected systems without permission and were subsequently denied access to the grid by the local utility company.

Permitting a PV system may take several months, so submit your application well in advance of the date you'd like to install the system, *and don't buy a PV system until your permit has been granted*. And, as a final note on the subject, keep a copy of the certificate of approval. You may need it in the future if you file an insurance claim.

Covenants and Neighborhood Concerns

Restrictive covenants can create a huge obstacle to PV system installations. Even with a permit from the local building department, restrictive covenants can block an installation.

Covenants are agreements incorporated into the document under which a subdivision was created which require or prohibit certain activities. These covenants "run with the land," meaning that they apply to subsequent owners of the property as well as the original owner. Some covenants give the neighborhood association the right to create and enforce additional rules and architectural standards. Restrictive covenants and neighborhood association rules may address many aspects of our lives — from the color of paint we can use on our homes to the installation of a privacy fence. Some neighborhood associations even ban clotheslines. Covenants and neighborhood association rules are found in many neighborhoods and subdivisions throughout North America. Some even expressly prohibit renewable energy systems, such as solar hot water systems and solar electric systems.

Restrictive covenants are legally enforceable and the courts have consistently upheld their legality. To see if you will be prohibited from installing a PV system, you need to review both the restrictive covenants and neighborhood association rules, if they exist. Contact your neighborhood association to see what rules apply.

If your subdivision has restrictive covenants or architectural standards, you need to apply for permission to install a PV system. The application is often a written letter with a drawing or two. Precedence can help. In other words, if someone else has installed a similar system, even without permission, it's easier to obtain approval.

Even in the absence of restrictive covenants, you might want to consider your neighbors. Many people feel very strongly about protecting the aesthetics of their neighborhoods. So, be sure to discuss your plans to install a PV with your closest neighbors, long before you lay your money down — unless you live out in the country.

Don't expect your neighbors to be as enthusiastic about a PV system on your rooftop or in your backyard as you are. Not everyone is enamored of renewable energy. Some people have a knee-jerk reaction against it. Why?

Many people fear the unknown. Others object to anything new or "environmental." Others may have distorted notions of what a solar system might look like. Still others may object out of spite or because of unresolved anger. And some just don't like the way a PV system looks.

To help allay fears and win over your neighbors, you may want to show your neighbors pictures of the PV system on your home. If you are good at Photoshop, take a photo of your home from their house and superimpose a picture of the PV system on your roof or on your property so they can see what it will look like. If you are thinking about installing a grid-connected system, let them know that you'll be supplying part of their energy, too.

Giving neighbors advance notice, answering questions they have, and being responsive to their concerns is the best way to ensure peace in your neighborhood. You may even consider installing a flush-mounted or direct-mounted PV system to avoid objections.

If you're building a new home in an historic district or adding a PV system to an existing home in an historic district, you may have a battle on your hands. A building-integrated PV system may be the only possible option, and even that may not meet approval. However, if you can locate the collector where it is not visible from the street, your chances for approval will be greater.

Although covenants can be a pain, some renewable energy-friendly states like Colorado and Wisconsin prohibit homeowners associations from preventing homeowners from installing solar and other renewable energy systems. If you're in one of those states, count yourself lucky. If not, you might want to work to pass such legislation in your state.

Connecting to the Grid: Working with Your Local Utility

When installing a grid-connected system, you'll need to consult with your local utility company (early on — *before* you purchase your equipment) to work out an arrangement to connect to their system. You'll need to file an application that provides sufficient detail to assure the utility that the electricity your PV system will be back feeding onto the grid will be of the same quality as grid power. They will also want assurance that your system won't continue to back feed electricity onto the grid if the grid goes down. They don't want their workers getting shocked by a PV system that's operating while their system is down. You'll also need to determine their policy on compensation for any surplus electricity you may be supplying to the grid.

While all this may sound daunting, bear in mind that small utility-interconnected PV systems are safe. As you learned in Chapter 6, the use of a grid-connected inverter ensures that their output matches grid power. A professional installer should be able to address the concerns of your utility and file the necessary paperwork.

As for payment, all utilities in the United States are required by federal law (Public Utility Regulatory Policy Act of 1978) to buy surplus electricity produced by their grid-connected customers. There's no getting around that. How much they pay is a different matter.

Federal law requires utilities to pay the *avoided cost* — that is, what it costs the utility to generate the power, which is usually one-fourth to one-third the amount they charge their customers. If you are paying 8 cents a kilowatt-hour, the avoided cost may be as low as 2 to 3 cents.

Fortunately, many states (41 at this writing) have enacted net metering policies that require utilities to pay residential clients the same price they charge. Under net metering, if they are charging customers 10 cents a kilowatt-hour, they credit producers like you the same rate for power you deliver to the grid until you start generating a surplus. How they deal with surpluses varies, as noted in Chapter 5.

Some utilities use existing electrical meters to keep track of the ebb and flow of electricity, provided they are bidirectional — capable of spinning backward and forward. (As noted in Chapter 5, these meters spin backward when electricity is being back fed onto the grid and forward when electricity is flowing from the grid to your home or business.) If your meter is unable to spin in two directions, the utility may require installation of a second meter, one to track the electricity you deliver to the grid. Or, they may require a new single digital meter that keeps track of the flow of electricity to and from the grid. If a new meter is required, you'll be expected to pay for it.

Some utilities also charge an interconnection fee. This fee may be based on the size of the system or may be fixed — with one price covering all systems. Fees typically range from $20 to $800. They cover costs borne by the utility, such as the cost of inspecting a PV system after it is installed. Be sure to ask about interconnection fees upfront and be sure the contract is clear about any additional fees that may be charged to you over the course of the contract. Also, be sure to ask whether the utility is claiming renewable energy certificates (RECs) for your PV system. RECs are financial instruments that can be sold to individuals wishing to offset their carbon emissions. They can be a valuable financial asset to some customers.

Getting Through to the Utility

When installing a grid-connected system, most installers contact the utility at the same time they file for a building permit. "Making initial contact with the utility could be one of your most formidable tasks," notes Mick Sagrillo in an article on utility requirements in *Windletter* published by the American Wind Energy Association. This is especially true in rural areas and in states that have only recently adopted net metering.

Unless your local utility has lots of experience with solar electric systems, they probably won't have a specific person who deals with interconnection agreements. If your utility is one of the more enlightened or experienced companies, however, they may have a point person who deals with such matters. Check out your utility's website and look for terms such as "net metering," "interconnection agreement," "renewable energy systems," or "distributed

generation." You may find all the information you need online.

Although utilities are required to allow small electrical generators to connect to the grid, getting them to cooperate may be another matter entirely. Problems are especially common in rural electric cooperatives, many of which have been hostile to the idea of allowing small-scale renewable energy generators to connect.

So, unless you know that your utility supports customer-owned renewable energy projects, call your state's *public utility commission* (PUC) before calling your utility. Their staff should be able to describe the approval process for the various types of utilities (publicly owned, rural electric, and municipally operated). They may also put you in touch with the appropriate contact person at the utility. "When you do finally contact your utility," Sagrillo notes, "it does not hurt to mention that you've been in touch with the PUC."

Working through the Details

"Once in contact with the utility, try to find out who is responsible for processing your interconnection application, and who will be making the final decision on your application," Sagrillo advises. Be prepared to answer all of that person's questions. If you do not know an answer to a question, admit it, and then find the answer yourself. Remember, your project is not their priority. If they don't know the answer, they're not likely to perform the necessary research, especially if they are opposed to such proposals. (Many companies own coal-fired power plants and see renewable energy as a threat.)

Before they sign a contract with you, the utility will very likely require you to submit a site plan indicating where components are located and electrical schematics. Its two most important concerns, however, will be an assurance that you have liability insurance and that your system includes an automatic disconnect. As noted above, all utility-connected inverters on the market incorporate an automatic disconnect feature. Even so, the utility may require an inspection of the inverter and even a demonstration.

When dealing with a utility that is inexperienced with grid connections, do not allow them to use their lack of familiarity with such systems as an excuse to deny you access, advises Sagrillo. "Nor should you allow them to charge you thousands of dollars in engineering fees to review your electrical schematics or test your system. Tens of thousands of PV and wind systems are connected to the grid across the US. This is not a new or unproven technology."

As noted in Chapter 5, utilities may also require installation of a visible, lockable disconnect — that is, a manually operated switch that allows utility workers to disconnect your system from outside your home in case the grid goes down and they need to work on the electrical lines. (Utilities may reserve the right to disconnect your PV system to work on their system without notifying you.) Because grid-connected inverters automatically disconnect from the grid when they sense a drop in line voltage or a change

in frequency, this requirement is unnecessary. Regardless, it may still be required by your local utility and will add to your installation costs.

Even though lockable disconnects are rarely, if ever, used, do not fight your utility on this requirement, says Mick. "In this and other situations, always remember: utilities often consider 'allowing' you to hook up your wind system to their grid analogous to 'doing you a favor,' even though they are required to by federal law."

Fostering a Productive Relationship

As when working with a building department, keep in mind that there is no bigger turn-off than having to deal with an overly assertive and arrogant know-it-all. While you may be in the right, the utility is a lot bigger than you are, with much more legal expertise and resources to fight your installation if they choose to. Your utility can make the approval process very fast or unbelievably difficult. "It has all the marbles, and it has the bag, too. And it knows it," Sagrillo notes. Few homeowners have the resources to challenge a utility in court. If you run into problems with the utility about any matter and feel you are right, your best bet is to work through your state's public utility commission.

If you go into the process armed with the proper knowledge, however, it is unlikely ever to come to that, notes Sagrillo. And how much better it is to invite your utility out to inspect your PV system, test the automatic shutdown features, and become a believer when the staff members actually witness for themselves your PV system generating electricity and feeding it onto the grid. Besides allowing a homeowner to interconnect his PV system with the grid, the utility will become better educated on home-sized PV systems and could become an advocate of the technology.

Making the Connection

Once a grid-connected PV system has passed final inspections by the local AHJ, you or the AHJ must contact the utility. At that point, they will sign the interconnection agreement, granting you the right to connect to the grid. You can flip the switch and start generating electricity. In some cases, however, utilities insist on inspecting and testing PV systems themselves to be sure the automatic disconnect functions as it is supposed to.

Insurance Requirements

Besides permits, inspections and utility approval for interconnection, if the system is grid connected, a property owner may also be required to secure insurance for a PV system. Two types of insurance are required: property damage which protects against *damage to the PV system* and liability insurance to protect you against *damage caused by the system*. Both are provided by homeowner's or property owner's insurance policies.

Insuring Against Property Damage

For homeowners, the most cost-effective way to insure a new PV system against damage is under an existing homeowner's insurance policy. (This is far cheaper than trying to secure a separate policy for a PV system.) Businesses can cover a PV system under their property insurance, too.

When installing a PV system on a home or outbuilding — or even on a pole mount — contact your insurance company to determine if your current coverage is sufficient. If not, you'll need to boost the coverage enough to cover replacement, including materials and labor. If you are installing a PV array on a pole, be sure to let them know that the PV system should be insured as an "appurtenant structure" on your current homeowner's policy. This is a term used by the insurance industry to refer to any uninhabited structure on your property not physically attached to your home. Examples include unattached garages, barns, sheds, satellite dishes and wind turbine towers.

Insurance premiums on a homeowner's insurance policy fall into two different categories, each with differing rates. Your home is assessed at a higher value than an unattached garage or a storage shed or your tower. This is because people's homes are more lavish than most garages and sheds, and contain a myriad of personal possessions, furniture and clothing not typically found in other structures.

Appurtenant structures, on the other hand, are assessed and charged at a lower rate. Insurance companies usually base premiums on appurtenant structures on the total cost of materials plus the labor to build the structure. Bear in mind that coverage of appurtenant structures on a homeowner's policy only applies if the system is installed on the same property as the house. If it is on a separate piece of property, it may have to be insured under a separate policy.

It is best, although a bit more expensive, to insure a PV system at its full replacement cost — not a depreciated value. PV systems can easily last two decades or more. Any PV system should have insurance coverage that includes damage to the system itself from "acts of nature," a.k.a. "acts of God," plus possible options for fire, theft, vandalism or flooding.

While most PV systems are designed to withstand 100-plus mile-per-hour winds, tornadoes or hurricanes can obviously destroy them, just as they would any other structure in their path.

Another "act of nature" of concern is lightning strikes. A properly installed PV system is properly grounded. Pole-mounted arrays are grounded at the base of the pole and the wire is grounded before it enters the home. Inside, the PV system ground is connected (bonded) to the electrical ground of the home's wiring. In addition, grid-connected systems are grounded on the utility side of the inverter.

Further protection is provided by installing lightning arrestors in appropriate locations. For example, a lightning arrestor should be installed in the combiner box and in the DC disconnect. A lightning arrestor and voltage surge arrestor should also be installed on the utility side of the inverter to provide additional protection.

While lightning arrestors will not guarantee that your system is safe from lightning,

it may reduce the damage from a rare direct or more likely nearby lightning strike. Plus, in the eyes of the insurance company, you have taken prudent measures to protect your system. If your system is fried by lightning, you'll be able to collect from the insurance company.

Fire is of minimal concern to a PV system. However, it should go without saying that the wiring of the entire system must comply with the electrical code. If you have a fire in the house caused by some bad wiring in the system, your claim may be denied and your policy canceled.

Theft of a PV system, or a part of the system, is unlikely, unless the modules are mounted on the ground. Vandalism may also be a concern for ground-mounted PV systems. While incidents of vandalism are infrequent, they have occurred.

Flood insurance is a nationally administered program to protect primary dwellings. Costs can be exceptionally high for a home located along a coastline or in a floodplain near a stream or river with a penchant for flooding. If you live in a floodplain, obtain an insurance estimate before beginning construction.

Insuring a PV system — either as an addition to your home or business or as an appurtenant structure on a homeowner's policy — is relatively inexpensive. While home insurance coverage should cover appurtenant structures, added insurance may be required. It can be purchased for an additional premium. Since most rural homeowner insurance claims are for fire damage, the deciding factor in pricing coverage is determined by the distance of the home or business from the nearest fire department.

Liability Coverage

Liability insurance, which is also part of homeowner's or business owner's insurance policy, protects against possible damage to others caused by a PV system. Although we have never heard of any such damage occurring, liability insurance is required by the utility for homeowners and businesses who wish to install a utility-connected PV system. Why is liability insurance required?

Liability insurance covers possible claims from damage to a neighbor's electronic devices from a grid-connected PV system. (For example, if your system sends power onto the line that somehow, magically, damages electronic equipment in a neighbor's home.) It also covers personal injury or death of employees due to electrical shock from a system when working on a utility line during a power outage. Even though the likelihood of this is nil, because of the automatic disconnect feature built into inverters, utilities still insist on this coverage. Don't resist them.

Liability coverage is relatively inexpensive and, as mentioned above, is part of a homeowner's or business owner's policies. "Liability insurance in the amount of $100,000 is considered adequate for small PV systems by most utilities," writes Jim Dunlop in his book, *Photovoltaic Systems*. In most places, liability coverage for homes runs from $100,000 to $300,000. Increasing liability coverage to $500,000 may add an additional $10 to an annual premium in most areas. Extending

coverage to $1 million may add $35 to $40 more to the annual premium.

The utility will likely dictate the level of liability insurance that it requires as a condition for interconnection. If the amount seems unreasonable to you, consider appealing to your state's public utility commission.

In addition to liability insurance, utilities may require you to indemnify them from potential damage from your PV system. Interconnection agreements may also limit the size of PV systems. Arizona, for instance, limits system size for net metering to 10 kW. Wisconsin set the limit at 20 kW. The state of Colorado, on the other hand, established a 2,000 kW limit. While limits such as those in Arizona and Wisconsin may seem restrictive, we should point out that 10- and 20-kW PV or PV and wind systems are larger than most families and many businesses will require. We should also point out that larger grid-connected systems can be installed, but are subject to more stringent regulations and may not qualify for net metering. Electricity from such systems is purchased at lower than retail rates, too. For more advice on insurance, and specifically on working with insurance agents, see the accompanying sidebar, "Insuring your PV System."

Buying a PV System

If a PV system seems like a good financial, social or recreational pursuit for you, we strongly recommend hiring a competent, bonded, trustworthy and experienced professional installer. A local supplier/installer with experience and knowledge can be a great ally. They can supply all of the equipment, be certain that it is compatible, obtain or assist with permits and interconnection agreements, and, then, of course, install the system. They will test it to be sure it is operating satisfactorily (don't pay in full until you are satisfied). They should be there to answer questions and to address problems you have with the equipment.

We recommend that you find an installer who stands solidly by his or her work. Most PV modules come with a 20- to 25-year production warranty, and grid-tied batteryless inverters are often guaranteed for five years. Many manufacturers offer extended warranties, most notably when a ten-year warranty is required to qualify for special installation

Insuring your PV System

Insuring a PV system should be pretty straightforward; however, not all insurance agents understand renewable energy systems. Ignorance of PV systems on the part of an insurance agent can be a significant obstacle. Be sure to contact your agent well before you buy your system.

When discussing this matter with your agent, explain things simply. Remember, they don't need all the details and probably wouldn't understand them. Meet in person if at all possible so you can clear up any misunderstandings. It's much easier to interpret confusion in a voice or face. Avoid e-mail communications.

Inform your agent that the solar electric system will be designed and installed by a professional according to the National Electrical Code and that all the equipment is UL listed. Also let the agent know that the design and installation will be approved by the building department and that it will be inspected as well for compliance with local codes.

You may want to take pictures and diagrams of PV systems, but use them only if your agent seems confused. Keep the diagrams simple. This should allay any concerns your agent might have. If you need help, ask your installer to talk with your agent. He or she can explain how safe PV systems are, and why additional liability insurance is not necessary.

Despite your efforts, some agents and insurance companies may still be unwilling to insure a PV system. If so, move on. Find an agent and company that is better informed.

When discussing insurance requirements, bear in mind that a PV system will add to the value of your home — so it may increase the cost of your homeowner's policy. Some agents may insist on raising the liability coverage, which could cost you several hundred dollars a year. Additional liability coverage was once required by utilities, but most state net metering laws prohibit this requirement. Let your agent know this.

Whatever you do, always be forthright and honest with your agent. Let them know the full cost of the system, including installation. Not informing an agent of your intent to install a PV system could come back to haunt you. If you file a claim at a later date, it is likely to be denied.

incentives. Battery-based inverters are often guaranteed for two to five years, but again, an extended warranty may be available. Reliable equipment and local installers who know the business are both worth their weight in gold. Look for people who've been in the business for a while and who have installed a lot of systems. The longer the better; the more systems the better.

Be sure your installer is bonded, too. A bond is sum of money set aside by contractors and held by a third party, known as a *surety company*. The bond provides financial recourse to homeowners and business owners if a contractor fails to meet his or her contractual obligations. In such instances, customers can file a claim for compensation from the bonding company. To stay bonded, a contractor must reimburse the company to cover any payments made to customers. Residential contractors usually carry a minimum of $10,000 bond and some AHJs require that contractors be bonded.

When contacting companies, ask if they are bonded and the amount of the bond. Check it out. Also ask for references, and be sure to call them. Visit installations, if possible, and contact the local office of the Better Business Bureau. Get everything in writing. Sign a contract. Be sure the installer has worker's compensation insurance to protect his or her employees when working on your site. Don't pay for the entire installation up front. Be sure you are on the site when the work is being done. The accompanying sidebar, "Questions to Ask Potential Installers," includes a list of questions you should ask when shopping for an experienced and competent installer.

You can also purchase equipment from a local supplier and install it yourself with or without their guidance, provided the AHJ allows property owners to perform electrical work. If you live in such an area, you can file for permits and install all equipment — PV arrays and electronics. However, you cannot legally hire any unlicensed electricians to act as the electrical contractor or employ any unlicensed individuals to work on the system. You have to do the work yourself, except for the final grid connection, which must be performed by a licensed electrical contractor. The AHJ may also require the owner/installer to live in or occupy the building and may restrict him from selling or leasing the property for one year.

As a rule, we don't recommend this route unless you are handy and have attended a couple of PV installation workshops. Installing a PV system is risky. Working on a roof and wiring are fraught with difficulties. Connecting to the electrical grid is a job for professionals. Even professional PV installers must hire licensed electricians to do the work. A PV system is a huge investment and you don't want to mess it up.

You can also purchase equipment online from manufacturers or from online dealers. Some vendors sell complete packages with PV modules, inverters, racks and everything else. Getting help installing packaged system from such suppliers can be difficult.

Another option is to buy from a local supplier or an online source and sponsor a workshop on your property to install the equipment. Organizations like The Evergreen Institute, Solar Energy International and the Midwest Renewable Energy Association are often looking for PV installations in different parts of the country. They fly in a PV expert who teaches a one- or two-week installation workshop. Workshop registrants pay a fee to cover the cost of the instructor, the costs of advertising and setting up a workshop, and perhaps even a portable toilet. The homeowner pays the costs of materials, and the workshop attendees provide much of the labor free.

Many installation workshops that Dan has attended in solar and wind had one to three licensed electricians participating as students in addition to the instructor; nonetheless, a local electrician may be required to oversee the installation. He or she would be the "installer of record," which is required for a permit.

Although you may not save any money on the installation, workshop installations can be very satisfying! You get a PV system installed on your home or business while providing a valuable opportunity for others to learn how

Questions to Ask Potential Installers

1. How long have you been in the business? (The longer the better.)
2. How many systems have you installed? How many systems like mine have you installed? (The more systems the better.)
3. How will you size my system?
4. Do you provide recommendations to make my home more energy efficient first? (As stressed in the text, energy efficiency measures reduce system size and can save you a fortune.)
5. Do you carry liability and worker's compensation insurance? Can I have the policy numbers and name of the insurance agents? (Liability insurance protects against damage to your property. Worker's compensation insurance protects you from injury claims by the installer's workers.)
6. Are you bonded? (Bonding, as explained in the text, provides homeowners with financial recourse if an installer does not meet his or her contractual obligations.)
7. What additional training have you undergone? When? Are you NABCEP certified? (Manufacturers often offer training on new equipment to keep installers up to date. NABCEP is a national certifying board that requires installers to pass a rigorous test and have a certain amount of experience.)
8. Will employees be working on the system? What training have they received? How many systems have they installed? Will you be working with the crew or overseeing their work? If you're overseeing the work, how often will you check up on them? If I have problems with any of your workers, will you respond immediately? (Be sure that the owner of the company will be actively involved in your system or that he or she sends an experienced crew to your site.)
9. Are you a licensed electrician or will a licensed electrician be working on the crew? (State regulations on who can install a PV system vary. ☞

to install systems. Some of them may become solar installers or start up solar businesses. You do, however, have to be comfortable with a dozen or more people on your site for a week. If the workshop leader is competent and checks all of the attendees' work, you'll get a PV system installed right. Be sure to contact the nonprofit organization well in advance and be prepared to help organize and publicize the workshop. Also, be sure your insurance will cover volunteer workers on your site.

Parting Thoughts

When you started reading this book, you no doubt were already interested in PV systems. Perhaps you just wanted to determine if a PV would be suitable for your home. Perhaps you were sure you wanted to install a PV system but didn't know how to proceed. We hope that you now have a clear understanding of what is involved.

If you have come to realize that your dreams for a PV system were not realistic, be glad that you did not spend a lot of money on

A licensed electrician may not be required, except to pull the permit, supervise the project, and make the final connection to your electrical panel.)

10. What brand modules and inverters will you use? Do you install UL listed components? (To meet code, all components must be UL listed or listed by some other similar organization.)

11. Do you guarantee your work? For how long? What does your guarantee cover? How quickly will you respond if troubles emerge? (You want an installer who guarantees the installation for a reasonable time and who will fix any problems that arise immediately.)

12. Do you offer service contracts? (Service contracts may be helpful early on to be sure the system runs flawlessly.) How much will a service contract cost? What does it cover?

13. Can I have a list of your last five projects with contact information? (Be sure to call references and talk with homeowners to see how well the installer performed and how easy he or she was to work with.)

14. What is the payment schedule? Can I withhold the final 10% of the payment for a week or two to be sure the system is operating correctly? (Don't pay for a system all at once. A deposit, followed by one or two payments protects you from being ripped off. Never make a final payment until you are certain the system is working well.)

15. Will you pull and pay for the permits?

16. Will you work with the utility to secure an interconnection agreement?

17. What's a realistic schedule? When can you obtain the equipment? When can you start work? How long will the whole project take?

■

a PV system that would not have met your expectations. Efficiency measures are a lot cheaper. If, however, you now have an informed conviction that a PV system is for you, you may want to locate a dealer/installer.

With inflation and the specter of rising costs of fuel and increasing demand for the materials that go into a PV system — aluminum, copper, silicon, concrete, etc. — prices of PV systems are not likely to come down. The longer you wait, the higher the cost of installing a system is likely to be. Also, the cost of electricity and the fuels needed to produce it are rising. The sooner you get your PV system in operation, the sooner you will start saving. If you are convinced that PVs are for you, and have a good site, you might want to get a move on.

You are now up to speed on photovoltaic systems. You know a great deal about solar energy, electricity, PV modules, PV systems, inverters, batteries, charge controllers and permits. You've gained good theoretical as well as practical information that puts you in good position to move forward. We've given you mountains of advice on installation and helped you grapple with economic issues. Our work has ended, but yours is just beginning.

We wish you the best of luck!

Resource Guide

Books

Amann, Jennifer, Alex Wilson and Katie Ackerly. *Consumer Guide to Home Energy Savings*, 9th ed., New Society, 2007. Excellent introduction to home energy efficiency.

Chiras, Dan. *The Homeowner's Guide to Renewable Energy*. New Society, 2006. Contains information on residential wind energy and solar electric systems.

Chiras, Dan. *Green Home Improvement: 65 Projects that Will Cut Utility Bills, Protect Your Health, and Help the Environment*. RS Means, 2008. Contains numerous simple projects to make your home more energy efficient and reduce your energy bills.

Dunlop, Jim. *Photovoltaic Systems*. American Technical Publishers, 2007. Superb resource for future installers and other solar energy professionals.

Ewing, Rex A. *Power With Nature: Solar and Wind Energy Demystified*. PixyJack Press, 2003. Skip to the second half if you want to get to the meat of the matter.

Ewing, Rex and Doug Hargrave. *Got Solar? Go Solar: Get Renewable Energy to Power Your Grid-Tied Home*, PixyJack Press, 2005. Great little book for those interested in installing a grid-tied PV system.

Komp, Richard J. *Practical Photovoltaics: Electricity from Solar Cells.* 3rd ed., Aatec Publications, 2002. A brief, but fairly technical book for more scientifically inclined readers.

Perlin, John. *From Space to Earth: The Story of Solar Electricity*. Aatec Publications, 1999. A detailed history of the development of PVs.

Schaeffer, John. *Solar Living Source Book*. New Society, 2008. Contains a lot of useful information on PV systems and PV products.

Scheckel, Paul. *The Home Energy Diet*. New Society, 2005. A great guide for energy conservation in homes.

Strong, Steven with William G. Scheller. *The Solar Electric House: Energy for the Environmentally Responsive, Energy-Independent Home*. Sustainability Press, 1993. An excellent resource but slightly out of date on some topics.

Solar Energy International. *Photovoltaics: Design and Installation Manual*. New Society, 2004. A great book for those interested in becoming an installer or individuals interested in installing their own PV systems.

Articles

Aldous, Scott, Zeke Yewdall and Sam Ley. "A Peek Inside a PV Cell," *Home Power* 121, 64-68, 2007. Describes the inner workings of a PV cell. Worth looking at if for nothing else but the excellent illustrations.

Bailes, Allison A. "How Efficient is Your House?" *Home Power* 106, 74-78, 2005. A great little piece on blower door tests, a simple, effective test designed to tell us how leaky our homes are and to identify leaks to reduce infiltration and exfiltration.

Chisholm, Grey. "Finding True South the Easy Way," *Home Power* 120, 64-66, 2007. How to use the night sky to determine true south.

Cohn, Lisa. "Safeguard Your RE Investment: Finding a Policy that Works for Your System," *Home Power* 128, 72-76, 2008. An insightful look at the insurance for RE systems.

Dankoff, Windy. "Lightning Happens," *Home Power* 107, 60-64, 2005. Important information on protecting a renewable energy system from lightning.

Dankoff, Windy. "Top Ten Battery Blunders," *Home Power* 115, 54-60, 2006. Important reading for anyone installing a battery bank.

Davidson, Kelly. "The Best States for Solar," *Home Power* 124, 84-90, 2008. Superb resource.

Donahue, Topher. "Solarscapes: A New Face for PV," *Home Power* 122, 36-40, 2007. Looks at an alternative way of mounting PVs to increase output and create shade.

Gudgel, Bob. "Get Maximum Power from Your Solar Panels with MPPT," *Home Power* 109, 58-62, 2005. Excellent article on how maximum power point tracking works.

Guevara-Stone, Laurie and Ian Woofenden. "Choosing Your RE Installer," *Home Power* 127, 48-52, 2008. A good guide for hiring a professional installer for a PV system.

Home Power Staff. "Clearing the Air: *Home Power* Dispels the Top RE Myths," *Home Power* 100, 32-38, 2004. Superb piece for those who want to learn about the many false assumptions and beliefs regarding renewable energy.

LaForge, Christopher. "Choosing the Best Batteries: 2009 Battery Specifications Guide," *Home Power* 127, 80-88, 2008. Excellent guide to lead acid batteries.

Livingston, Phil. "First Steps in Renewable Energy," *Home Power* 118, 68-71, 2007. A good primer on home energy savings, vital to slashing the cost of a PV system.

Marken, Chuck and Justine Sanchez. "PV vs. Solar Water Heating: Simple Solar Payback," *Home Power* 127, 40-45, 2008. Intriguing look at the economics of these two solar systems.

Mayfield, Ryan. "Grid-Direct PV Systems," *Home Power* 126, 50-51. A good introduction to grid-connected PV systems.

Mayfield, Ryan. "Rack and Stack: PV Array Mounting Options," *Home Power* 124, 58-64. Great discussion of PV mounts.

Mayfield, Ryan. "String Theory: PV Array Voltage Calculations," *Home Power* 125, 66-70. Very important article on sizing PV strings to match inverters and climate.

Mayfield, Ryan. "The Whole Picture: Computer- Based Solutions for PV System Monitoring," *Home Power* 121, 102-108. Very important discussion of ways to monitor PV systems, including web-based systems.

Meyer, John and Joe Schwartz. "Battery Box Basics," *Home Power* 119, 50-55, 2007. Superb reference for individuals who want to build their own battery boxes.

Munro, Khanti. "Battery Monitoring," *Home Power* 128, 92-94, 2009. A general overview of battery monitoring hardware.

Pahl, Greg. "Choosing a Backup Generator," *Mother Earth News* 202, 38-43, 2004. A fairly detailed overview of what to look for when buying a back-up electrical generator for a RE system.

Parker, Tehri. "Choosing the Right Site to Maximize your Solar Investment," *Home Power* 115, 30-33, 2006. A great description of how the Solar Pathfinder is used to assess solar resources and shading.

Perez, Richard. "Flooded Lead-Acid Battery Maintenance," *Home Power* 98, 76-79, 2004. Read and memorize this and put the advice into practice if you plan on installing a stand-alone RE system.

Perez, Richard. "Off-Grid Appliances: AC or DC?," *Home Power* 115, 52-53, 2006. A frank and useful discussion of DC appliances for anyone thinking of installing a DC only or DC circuits to bypass the inverter of their off-grid home to save energy.

Perez, Richard. "Off-Grid Inverter Efficiency," *Home Power* 113, 36-37, 2006. A brief, but important overview of inverter efficiency that could help off-gridders get the most from the PV systems.

Perez, Richard. "To Track or Not to Track," *Home Power* 101, 60-63, 2004. A great overview of tracker technologies and their pros and cons.

Pinkham, Linda. "From the Ground Up: My RE System Design Choice," *Home Power* 106, 44-49, 2005. Great case study for those who are contemplating installing a PV system; covers many basic questions you'll need to answer.

Root, Benjamin. "Deciphering Schematics," *Home Power* 123, 44-46. An excellent article for those who are interested in installing their own system or becoming an installer.

Russell, Scott. "Making Sense of Solar Electric System Cost," *Home Power* 109,

22-26, 2005. An excellent article that shows how to calculate the cost of a solar electricity system.

Russell, Scott. "Solar-Electric Systems Simplified," *Home Power* 104, 72-78, 2004. A well-illustrated overview of the types of PV systems and their components.

Russell, Scott. "Starting Smart: Calculating Your Energy Appetite," *Home Power* 102, 70-74, 2004. Great introduction to household load analysis (to determine household electrical demand).

Sanchez, Justine. "PV Energy Payback," *Home Power* 127, 32-36, 2008. Shows that PV systems produce as much energy as is required to make them fairly quickly.

Sanchez, Justine. "PV Module Buyer's Guide," *Home Power* 128, 78-89, 2008. Describes PV module specifications and lists specifications for modules on the market today.

Sagrillo, Mick. "Payback: The Wrong Question," *Windletter* 26(7), 1-3, 2007 Examines the topic of payback, especially why it is the wrong way to think about a renewable energy system.

Sagrillo, Mick. "Planning Your Wind System — Utility Requirements," *Windletter* 25(8), 1-3, 2006. Sound advice on working with the utility when installing a grid-connected wind system that's also relevant to PV systems.

Sagrillo, Mick. "Planning Your Wind System — Utility Buy-Back Rates," *Windletter* 25(9), 2006: 1-3. Important discussion of buy-back rates.

Schwartz, Joe. "DP&W Power Rail PV Mounts," *Home Power* 104, 82-84, 2004.

Describes Direct Power and Water Corporation's roof mount system.

Schwartz, Joe. "Finding the Phantoms," *Home Power* 117, 64-67, 2007. Excellent reading.

Schwartz, Joe. "How to Install a Pole-Mounted Solar-Electric Array, Part 2," *Home Power* 109, 82-89, 2005. Extremely well-written and well-illustrated guide on installing a pole-mounted PV array.

Schwartz, Joe. "Solar Electric Tools of the Trade," *Home Power* 105, 22-26, 2004. Very well illustrated guide for do-it-yourselfers or individuals interested in becoming a PV installer.

Schwartz, Joe. "Solmetric SunEye," *Home Power* 121, 88-90, 2007. A detailed look at a valuable solar site assessment tool.

Schwartz, Joe. "What's Going On — The Grid?" *Home Power* 106, 26-32, 2005. Excellent look at many of the grid-tied inverters.

Schwartz, Joe. "Wiley Electronics' ASSET," *Home Power* 119, 74-77, 2007. A detailed look at another valuable solar site assessment tool.

Schwartz, Joe and Zeke Yewdall. "Under Control: Charge Controllers for Whole-House Systems, *Home Power* 116, 80-84, 2007. Great overview of the various charge controllers on today's market.

Scheckel, Paul. "Efficiency Details for a Clean Energy Change," *Home Power* 121, 40-45, 2007. A quick guide to the ten most important energy efficiency measures.

Sharp, Jon, Ray Furse and Robert Chew. "Solar Success in the Northeast," *Home Power* 106, 50-54, 2007. An excellent article on economic incentives for those living in the northeastern United States.

Sindelar, Allan and Phil Campbell-Graves. "How to Finance Your Renewable Energy Home," *Home Power* 103, 94-99, 2004. Very useful article.

Sklar, Scott. "Solar-Electric Modules: Clean Energy· from Cradle to Cradle," *Home Power* 116, 34-36, 2007. An in-depth but slightly outdated look at the energy and environmental impacts of manufacturing PV cells.

Stone, Laurie. "Hiring a PV Pro," *Home Power* 114, 48-51, 2006. Great article with an especially interesting sidebar on things an installer won't do for you.

Swezey, Blair and Lori Bird. "Buying Green Power — You Really Can Make a Difference," *Solar Today* 17(1), 28-31, 2003. An in-depth look at ways home-owners can tap into renewable energy (such as green tags) without installing a system on their home.

Taylor, Jeremy. "Pump Up the Power: Getting More from Your Grid-Tied PV System," *Home Power* 127, 72-77, 2008. Good advice on ways to increase the efficiency of a PV system.

Thurston, Charles W. "Financing the Solar Dream; Leases and Power Purchase Agreements," *Home Power* 128, 52-56, 2008. An intriguing look at alternative ways to get a PV system on your home or business.

Truog, Jeremy. "Simple Steps to Save Through the Seasons," *Home Power* 115, 64-67, 2006. Excellent advice on saving energy in your home to reduce your PV system size and save money.

Weliczko, Erika. "Crystalline vs. Thin-Film," *Home Power* 127, 98-101, 2008. In-depth look at the differences between two module types.

Witte, John. "Install a Solar-Electric Roof," *Home Power* 114, 16-19, 2006. Great description of what's involved when installing solar shingles.

Woofenden, Ian. "Battery Filling Systems of the Americas: Single-Point Watering System," *Home Power* 100, 82-84, 2004. This article is a must for those who would like to reduce battery maintenance.

Woofenden, Ian. "Off or On Grid? Getting Real," *Home Power* 128, 40-45, 2008. An in-depth look at the pros and cons of grid-connected and off-grid PV systems.

Woofenden, Ian and Chris LaForge. "Getting Started with Renewable Energy: Professional Load Analysis and Site Survey," *Home Power* 120, 44-47, 2007. A good article for those who are thinking about having a renewable energy system installed.

Magazines

BackHome. PO Box 70, Hendersonville, NC 28793. Tel: (800) 992-2546. Website: BackHomeMagazine.com. Publishes articles on renewable energy and many other subjects for those interested in creating a more self-sufficient lifestyle.

Backwoods Home Magazine. P.O. Box 712, Gold Beach, OR 97444. Tel: (800) 835-2418. Website: backwoodshome.com.

Publishes articles on all aspects of self-reliant living, including renewable energy strategies such as solar.

The CADDET Renewable Energy Newsletter. 168 Harwell, Oxfordshire OX11 ORA, United Kingdom. Tel: +44 123335 432968. Quarterly magazine published by the CADDET Centre for Renewable Energy. Covers a wide range of renewable energy topics.

EREN Network News. Newsletter of the Department of Energy's Energy-Efficiency and Renewable Energy Network. See listing under organizations.

Home Energy Magazine. 2124 Kittredge Street, No. 95, Berkeley, CA 94704. Great resource for those who want to learn more about ways to save energy in conventional home construction.

Home Power. P.O. Box 520, Ashland, OR 97520. Tel: (800) 707-6585. Website: homepower.com Publishes numerous extremely valuable how-to and general articles on renewable energy, including PVs, wind energy, micro hydro, solar hot water and passive solar heating and cooling. This magazine is a goldmine of information, an absolute must for anyone interested in learning more. Magazine also contains important product reviews and ads for companies and professional installers. CDs containing back issues can be purchased from *Home Power.*

Mother EarthNews, 1503 SW 42nd St., Topeka, KS 66609. Website: motherearthnews.com. One of my favorite magazines. Usually publishes a very useful article in each issue on some aspect of renewable energy.

Solar Today. ASES, 2400 Central Ave., Suite G-1, Boulder, CO 80301. Tel: (303) 443-3130. Website: ases.org/. This magazine published by the American Solar Energy Society contains lots of good information on passive solar, solar thermal, photovoltaics, hydrogen and other topics, but not much how-to infromation. Also lists names of engineers, builders and installers and lists workshops and conferences.

Videos

An Introduction to Storage Batteries for Renewable Energy Systems with Richard Perez. This is one of the best videos in the series. It's full of great information. Produced by Scott S. Andrews, P.O. Box 3027, Sausalito, CA 94965. Tel: (415) 332-5191.

Organizations

Centre for Alternative Technology. Address: Machynlleth, Powys SY20 9Az. Tel: 01654 703409. Website: cat.org.uk/. This educational group in the United Kingdom offers workshops on alternative energy, including wind, solar and micro-hydroelectric.

Energy Efficiency and Renewable Energy Clearinghouse. P.O. Box 3048, Merrifield, VA 22116. Tel: (800) 363-3732. Great source of a variety of useful information on energy efficiency.

Energy Efficient Building Association. 490 Concordia Ave., P.O. Box 22307, Eagen, MN 55122. Tel: (651) 268-7585.

Website: eeba.org/. Offers conferences, workshops, publications and an online bookstore.

The Evergreen Institute, Center for Renewable Energy and Green Building. 9124 Armadillo Trail, Evergreen, CO 80439. Tel: (303) 883-8290. Website: danchiras.com. Teaches workshops on solar electricity, wind energy, passive solar heating and cooling, energy efficiency and much more through their educational center in Gerald, Missouri.

Real Goods Solar Living Institute. P.O. Box 836, Hopland, CA 95449. Tel: (707) 744-2017. Website: solarliving.org. A nonprofit organization that offers frequent hands-on workshops on solar and wind energy and many other topics.

Solar Energy International. Contact them at P.O. Box 715, Carbondale, CO 81623. Tel: (970) 963-8855. Website: solarenergy.org. Offers a wide range of workshops on solar energy, wind energy and natural building.

Solar Living Institute. P.O. Box 836, Hopland, CA 95449. Tel: (707) 744-2017. Website: solarliving.org. A nonprofit organization that offers frequent hands-on workshops on solar energy and many other topics. Be sure to tour their facility if you are in the "neighborhood."

The Evergreen Institute, Center for Renewable Energy and Green Building. 3028 Pin Oak Road, Gerald, MO 63037. Tel: (303) 883-8290. Website: evergreeninstitute.org. We offer hands-on workshops in residential energy efficiency, solar electricity, passive solar, wind energy, natural building and green building in Missouri and Colorado.

US Department of Energy and Environmental Protection Agency's ENERGY STAR program. Tel: (888) 782-7937. Website: energystar.gov.

PV Manufacturers

Advent solar — adventsolar.com
BP Solar — bpsolar.com
Canadian Solar, Inc. — csisolar.com
Day4Energy — day4energy.com
Evergreen — evergreensolar.com
GE — gepower.com/solar
Kaneka — kaneka.com
Kyocera — kyocerasolar.com
Mitsubishi — mitsubishielectric.com /searchsolar
Sanyo — sanyo.com
Schott — us.schott.com
Schuco — schuco-usa.com
Sharp — solar.sharpusa.com
SolarWorld — solarworld-usa.com
SunPower — sunpowercorp.com
Suntech Power — suntech-power.com
Sunwize — sunwize.com
Unisolar — unisolar.com
Yingli — yinglisolar.com

Inverter Manufacturers

Beacon Power Corp — beaconpower.com
Fronius USA — fronius.com
Magnetek, Inc. — alternative-energies.com
Outback Power Systems — outbackpower.com
PV Powered LLC — pvpowered.com

Sharp Electronics — sharp-usa.com/solar
SMA America, Inc. — sma-america.com
Xantrex Technology, Inc. — xantrex.com

Charge Controllers

Apollo Solar — apollo-solar.net
Blue Sky Energy, Inc. — blueskyenergy-inc.com
MidNite Solar, Inc. — midnitesolar.com
Morningstar Corp. — morningstarcorp.com
OutBack Power Systems — outbackpower.com
Xantrex Technology, Inc. — xantrex.com

Battery Manufacturers
Flooded Lead Acid

Deka/MK — eastpenndeka.com
Exide Battery — hawkerpowersource.com
GB HUP — enersysmp.com
Power Battery — powerbattery.com
Surrette/Rolls Battery — surrett.com
Trojan Battery — trojanbattery.com
US Battery — usbattery.com

Sealed Batteries

Concorde Battery — concordebattery.com
Deka/MK — eastpenndeka.com

Discover Energy — discoverenergy.com
Exide — exide.com
Exide Sonnenschein — exide.com
FullRiver — fullriver.com
Hawker — hawkerpowersource.com
Power Battery — powerbattery.com
Trojan — trojanbattery.com

Rack Manufacturers

Conergy — conergy.us
Direct Power and Water — directpower.com
Lorentz — lorentz.de
ProSolar — prosolar.com
Schuco — schuco-usa.com
Sharp — sharpsolaritson.com
Solar Earth — sunearthinc.com
SunPower — sunpowercorp.com
UniRac — unirac.com
Wattsun — wattsun.com
Zomeworks — zomeworks.com

Battery Monitor Manufacturers

Bogart Engineering — bogartengineering.com
OutBack Power Systems — outbackpower.com
Xantrex — xantrex.com

Notes

Chapter 1

1. The modules are not mounted directly on the roof, but on a metal rack that creates an air space behind the modules so they don't get hot.

Chapter 2

1. A small portion of the Sun's daily output consists of subatomic particles. This is known as *corpuscular radiation*. It consists primarily of beta particles (electrons) and alpha particles (protons and neutrons that are bound together). They're responsible for ionizing gases in the atmosphere, producing the Northern Lights.

Chapter 3

1. Efficiency in solar cells is a measure of the amount of electrical energy produced from a given amount of light. A 4 percent efficient cell converts 4 percent of the energy in sunlight into electrical energy.

2. Hybrid PV cells, discussed later in the chapter, have cell efficiencies ranging from 18 to 22%.

3. Approximately 35% of the incident rays would be reflected off the surface of a module without the anti-reflective coating. Multilayer coatings reduce reflection loss to less than 3%.

Chapter 5

1. In the state of Washington, grid-connected systems with battery backup have three meters if the owner wants to apply for a production incentive (a financial incentive based on electrical production of the system). This requirement prevents clients from buying energy at lower rates to charge batteries and then selling electricity from their batteries at higher "solar" rates.

2. In hot climates, where air conditioning is used heavily in the summer, surplus production may occur in the spring and fall months, and net consumption occur in the summer and winter months. But the advantage of net metering remains the same — surpluses can be banked and used later when needed.

3. A UPS large enough to supply a whole house would be cost prohibitive, although computer data centers and other highly critical applications have them.

4. An amp is 6.23×10^{18} electrons passing a point on a conductor per second.

5. In fact, it is important to program the set points of the charge controller higher than the values of the inverter, so that the charge controller doesn't try to regulate the batteries unless there is an outage. If a mistake were made and the charge controller's set point was set to float the battery a tenth of a volt lower than the float voltage setting of the inverter, you'd never send a single watt-hour to the home or grid because the charge controller wouldn't allow the battery voltage to come up to the "sell" point of the inverter (its float setting). The charge controller setting is typically a tenth volt higher, and if there is a temperature compensation sensor on the inverter, it has to be on the charge controller, too.

6. Phase 1 EPA emissions regulations for small spark ignition engines of 25 HP or less went into effect September 1, 1997 and reduce hydrocarbon emissions by about one-third. Phase 2 regulations would reduce emissions by another 35% and will go into effect in 2011 and 2012. There are also EPA emissions regulations for small diesel engines.

Chapter 6

1. Some inverters increase voltage without the use of a transformer. They use electronic voltage to boost circuits.

2. The voltage step-up or step-down ratio is proportional to the turns ratio.

3. The user must program the inverter with the local cost per kilowatt-hour and pounds of CO_2 per kilowatt-hour of utility supplied electricity.

Chapter 7

1. L16 is a standard battery size designation that is commonly used in renewable energy systems.

2. It is important that the charge controller can be — and is — configured for sealed batteries.

3. Regardless of how the batteries are connected, the total energy (watt-hours) stored is the same. Total watt-hours is determined by the watt-hour capacity of each battery and the number of batteries used. It would be clearer if batteries carried a watt-hour rating, since we use watts and watt-hours in most other aspects of system design. Watts (rate of energy generation, transfer or use) and watt-hours (quantity of energy generated, transferred or used) are easier to understand than amps (rate of charge flow) and amp-hours (quantity of charge moved).

4. According to Chris LaForge, an experienced PV installer who teaches workshops on PV installation for the Midwest Renewable Energy Association, it's generally best not to install more than three strings in parallel.

5. The idea that a concrete floor will "draw" charge out of a battery is a myth. The battery case is an insulator and no current will flow through the case regardless of what the battery is sitting on. Two good reasons for not sitting a battery on concrete are (1) a concrete floor is generally cooler than the surrounding room and the low temperature may not be good for the battery, and (2) acid mist from the battery will etch the floor and leave permanent marks on it.

6. Electrocution is not a hazard at 12, 24 or 48 volts. Dropping a tool or other metal object across the battery terminals could result in an electrical arc that can cause burns, or could result in an explosion of a battery case resulting in acid burns. Unprotected batteries are dangerous and should be located in secure containers or spaces.

7. The amp-hour capacity is based on a specified discharge rate, typically 8, 10 or 20 hours. If a battery is discharged at the specified rate, it should yield the specified, or rated, amp-hours. If it is discharged at a higher rate, it will yield less electricity.

8. Distilling and deionizing are different processes that produce similar quality pure water. Water from a reverse osmosis system is also suitable for batteries, provided that the system is working, i.e., the total dissolved solids (TDS) has been tested recently and is low.

9. There are some highly specialized PV systems where a charge controller may not be required — specifically when the array's peak current output (multiplied by 1 hour) is less than 3% of the battery amp-hour capacity. However, this applies to systems for navigational lights and other specialized applications, not to homes or businesses.

10. Some companies set the voltage points in their controllers at the factory and such settings are non-adjustable, other controllers can be adjusted by the user.

11. A word of caution. Check with the dealer or manufacturer of your gen-set or its engine before using biodiesel. It may void the warranty.

Chapter 8

1. Because an array stops producing DC electricity at the end of a day, running a tracker off DC directly from the array is not recommended. The array will not return to the east. For the second option, DC electricity from a 48-volt battery bank must be converted to 24-volt DC to power the electric motors that control the array; unfortunately, the DC-to-DC converters are not very reliable. As a result, most installers operate array-tracking motors on AC electricity. This requires an AC-to-DC power converter, which is a reliable device.

2. The length of this path varies with latitude. At 50° north (southern Canada),

the path on June 21, the summer solstice, is about 260°. At 40° north (Denver, CO, Columbus, OH and New York, NY) the path is about 244°. At 30° north (Houston, TX and St. Augustine, FL) the path is about 230°.

3. While the tilt angle is often set at the latitude of a site, NASA data available on the Surface Meteorology and Solar Energy website often indicates that the optimum angle is slightly different. Be sure to check the NASA charts before installing an array.

Index

Dan Chiras.

About the Authors

Dan Chiras is an internationally acclaimed author who has published over 24 books, including *The Homeowner's Guide to Renewable Energy* and *Green Home Improvement*. He is a certified wind site assessor and has installed several residential wind systems. Dan is director of The Evergreen Institute's Center for Renewable Energy and Green Building (www.evergreeninstitute.org) in east-central Missouri where teaches workshops on small wind energy systems, solar electricity, passive solar design, and green building. Dan also has an active consulting business, Sustainable Systems Design (www.danchiras.com) and has consulted on numerous projects in North America and Central America in the past ten years. Dan lives in a passive solar home powered by wind and solar electricity in Evergreen, Colorado.

Robert Aram is a renewable energy advocate and consultant. His consulting projects include off-grid farmsteads, highly energy-efficient homes, and a solar system for a university sports complex. He has participated in the design and installation of photovoltaic, solar thermal, and small wind systems. He is a registered professional engineer in Indiana and a Leadership in Energy and Environmental Design Accredited Professional.

Kurt Nelson has been installing solar electric systems through his company SOLutions since the late 1980s. He specializes in off-grid PV and teaches classes for the Midwest Renewable Energy Association.

If you have enjoyed *Power from the Sun* you might also enjoy other

BOOKS TO BUILD A NEW SOCIETY

Our books provide positive solutions for people who want to
make a difference. We specialize in:

**Sustainable Living • Green Building • Peak Oil • Renewable Energy
Environment & Economy • Natural Building & Appropriate Technology
Progressive Leadership • Resistance and Community
Educational and Parenting Resources**

New Society Publishers

ENVIRONMENTAL BENEFITS STATEMENT

New Society Publishers has chosen to produce this book on Enviro 100, recycled paper made with **100% post consumer waste**, processed chlorine free, and old growth free.

For every 5,000 books printed, New Society saves the following resources:[1]

35	Trees
3,189	Pounds of Solid Waste
3,508	Gallons of Water
4,576	Kilowatt Hours of Electricity
5,797	Pounds of Greenhouse Gases
25	Pounds of HAPs, VOCs, and AOX Combined
9	Cubic Yards of Landfill Space

[1]Environmental benefits are calculated based on research done by the Environmental Defense Fund and other members of the Paper Task Force who study the environmental impacts of the paper industry.

For a full list of NSP's titles, please call 1-800-567-6772 *or check out our website at:*

www.newsociety.com

NEW SOCIETY PUBLISHERS